AF864238

Coastal Research Library

Volume 37

Series Editor
Charles W. Finkl, Department of Geosciences, Florida Atlantic University, Boca Raton, FL, USA

The aim of this book series is to disseminate information to the coastal research community. The Series covers all aspects of coastal research including but not limited to relevant aspects of geological sciences, biology (incl. ecology and coastal marine ecosystems), geomorphology (physical geography), climate, littoral oceanography, coastal hydraulics, environmental (resource) management, engineering, and remote sensing. Policy, coastal law, and relevant issues such as conflict resolution and risk management would also be covered by the Series. The scope of the Series is broad and with a unique cross-disciplinary nature. The Series would tend to focus on topics that are of current interest and which carry some import as opposed to traditional titles that are esoteric and non-controversial.

Monographs as well as contributed volumes are welcomed.

Charles W. Finkl, Coastal Education & Research Foundation, USA Email: cfinkl@cerf-jcr.com

More information about this series at http://www.springer.com/series/8795

Risa Palm • Toby Bolsen

Housing Market Response to Sea-Level Rise in Florida

Springer

Risa Palm
Urban Studies Institute
Georgia State University
Atlanta, GA, USA

Toby Bolsen
Political Science Department
Georgia State University
Atlanta, GA, USA

ISSN 2211-0577 ISSN 2211-0585 (electronic)
Coastal Research Library
ISBN 978-3-030-88434-5 ISBN 978-3-030-88435-2 (eBook)
https://doi.org/10.1007/978-3-030-88435-2

© The Editor(s) (if applicable) and The Author(s), under exclusive license to Springer Nature Switzerland AG 2022
This work is subject to copyright. All rights are solely and exclusively licensed by the Publisher, whether the whole or part of the material is concerned, specifically the rights of translation, reprinting, reuse of illustrations, recitation, broadcasting, reproduction on microfilms or in any other physical way, and transmission or information storage and retrieval, electronic adaptation, computer software, or by similar or dissimilar methodology now known or hereafter developed.
The use of general descriptive names, registered names, trademarks, service marks, etc. in this publication does not imply, even in the absence of a specific statement, that such names are exempt from the relevant protective laws and regulations and therefore free for general use.
The publisher, the authors, and the editors are safe to assume that the advice and information in this book are believed to be true and accurate at the date of publication. Neither the publisher nor the authors or the editors give a warranty, expressed or implied, with respect to the material contained herein or for any errors or omissions that may have been made. The publisher remains neutral with regard to jurisdictional claims in published maps and institutional affiliations.

This Springer imprint is published by the registered company Springer Nature Switzerland AG
The registered company address is: Gewerbestrasse 11, 6330 Cham, Switzerland

To Eliot, Emma, Justin, and Katelyn

Acknowledgments

The authors would like to thank Georgia State University for providing funding to support graduate student assistants and the costs of acquiring survey responses. We thank Justin Kingsland, December Weir, and Chelsea Johnson for their help in data collection, cleaning, analysis, text editing, and the creation of several of the figures in this book. We appreciate Laura Villo's assistance in handling the Amazon gift cards for the real estate agent participants. We thank Jeremy Craig for his careful review of the manuscript and his suggested changes. We thank the First Street Foundation for permission to use images from Flood Factor. We are grateful for the helpful suggestions from several colleagues including Donald Reitzes, Jan Nijman, Ann-Margaret Esnard, Nancy Marlin, and David Greenland. Jeremy Kopkas provided both editorial advice and prepared the manuscript following the publisher's guidelines for electronic submission. Of course, any errors should be attributed to the authors.

Contents

List of Figures

List of Tables

About the Authors

Risa Palm is the author of 14 books and monographs as well as numerous scholarly articles and invited book chapters. Among her previous books are textbooks for introductory human geography courses (*An Invitation to Geography*, McGraw-Hill, 2 editions), upper division courses in American cities (*Geography of American Cities*, Oxford University Press), and natural hazards (*Natural Hazards: An Integrative Framework for Research and Planning*, Johns Hopkins University Press). She has also published three monographs on earthquake hazards with Westview Press (now part of Harper Collins) including *Earthquake Insurance: A longitudinal study*, *Earthquake Insurance in California: Environmental policy and Individual Decision-Making*, and Illusions of Safety: Culture and Earthquake Hazard Response in the US and Japan, and another monograph on earthquake hazards with The University of Chicago Press Geography Research Papers (*After a California Earthquake: Attitude and Behavior Change*). She has received research honors from the American Association of Geographers and served as president of this association.

Toby Bolsen is the author of over 40 scholarly articles and book chapters. His research focuses on the study of political behavior, public opinion formation, political communication, experimental methods, and US energy and climate policy. He has published research in numerous high-impact journals such as the *American Journal of Political Science*, *Political Behavior*, *Public Opinion Quarterly*, *Journal of Communication*, *Political Communication*, *Climatic Change*, *Science Communication*, *Energy Policy*, *Environmental Communication*, and many other outlets. He has received several best paper awards for published work presented at academic conferences.

Chapter 1
Introduction

Abstract Despite its vulnerability to flooding associated with sea-level rise, coastal South Florida continues to attract large numbers of new residents each day. Information about flood insurance requirements and flood-risk is widely available, but our 2018 survey found that coastal residents seemed relatively unconcerned about the future value of their homes. At the same time, some econometric studies have found that proximity to the coast may have a negative impact on both demand and sales price of coastal housing. Clearly, the south Florida coast is both an amenity and increasingly a hazard. Two other factors potentially impacted the market as of 2020. First, as of the summer of 2020, detailed, property-level information was made available about flood risk to prospective homebuyers on the website of the National Association of Realtors®. Second, the price and availability of flood insurance came into greater focus as there was serious discussion of revising the price structure of federal flood insurance for properties at substantial risk. To gain a perspective on the impacts of increased attention to the hazard of sea-level rise on the residential property market, we administered a survey in the fall of 2020 to 680 licensed Florida estate brokers and sales associates. We asked the real estate agents to describe what they observed in working with prospective south Florida homebuyers, and their assessment of local housing market trends. The analysis of their views has important implications for understanding how the risks of climate change and sea-level rise are reflected in the housing market both now and in the near-term future.

Keywords Real estate agents · Sea-level rise · Flood risk · Florida

Florida real estate. What image does this term suggest? Is it life on the beach with palm trees swaying and summer weather all year round (Fig. 1.1)? Is it the view of the ocean from a condominium window or the feel of pleasant breezes while walking along an oceanfront boardwalk? Is a Florida home on the coast one of your dreams?

Or do you think about the hurricanes that threaten Florida every summer and fall? Or the photos of the streets of Miami on a sunny day flooded by tidal waters? Or the summer days when the heat and humidity seem unbearable? Or do you think about

© The Author(s), under exclusive license to Springer Nature Switzerland AG 2022

R. Palm, T. Bolsen, *Housing Market Response to Sea-Level Rise in Florida*, Coastal Research Library 37, https://doi.org/10.1007/978-3-030-88435-2_1

Fig. 1.1 Florida beach. (Source: iStock photo)

the days when slick salespersons sold swampland to hapless buyers from out-of-state?

This book is about the south Florida residential real estate market and its vulnerability to sea-level rise resulting from climate change. We will explore real estate agent opinions on how prospective homebuyers assess the risk of flooding, and whether they observe that house prices are stagnant or falling in coastal areas vulnerable to flooding. And we will report on their conclusions after working with prospective homebuyers as to whether coastal south Florida is a good place to find a home or, alternatively, a risky investment in a place that will eventually be submerged by rising seas.

The question of how sea-level rise is affecting the south Florida property market is an urgent one: the residential real estate market is one of the pillars of Florida's economy, accounting for 22% of the Florida gross domestic product or over $240 billion annually (Woetzel et al. 2020). If prices fall precipitously, households might be unable to obtain mortgage financing or affordable insurance. A decline in property values would mean losses of local property tax revenues, with concomitant deterioration of support for regional infrastructure and education. At the national and international level, a rapid decline in the value of coastal Florida property would have a jarring and cascading effect throughout the insurance and secondary mortgage markets as well as the variety of entities that have financial stakes in banking and real estate in Florida (Nicholson et al. 2018; Taylor 2020; Taylor and Weinkle 2020).

Precise estimates of the scale and timing of sea-level rise for the Florida coast, or indeed any location, remain elusive (Hinkel et al. 2019; Kopp et al. 2019).

Assessments of local vulnerability require information on several parameters not currently known, including the degree to which societies change rates of carbon dioxide emissions, and uncertainty about the dynamics of Antarctic ice sheet, particularly the collapse of ice-cliffs (Pattyn and Morlighem 2020). Locally, estimates of sea-level rise also require information on the local and regional impacts of current and future human-built structures such as seawalls, beach nourishment or groins, as well as information about on wave movement and rates of erosion (Nordstrom, 1994), and rates of future local land subsidence (Wdowinski et al. 2020). Because of these uncertainties, climate science can at best provide scenarios and probabilistic estimates (Jackson and Jevrejeva 2016; Jevrejeva et al. 2019). However, despite these sources of uncertainty in the science, the consensus conclusion for the Florida coast is clear: the pace of sea-level rise is accelerating, and billions of dollars of property are at risk (Elliott 2021; Fu et al. 2016; Ghanbari et al. 2020; Armal et al. 2020).

At issue is the question of whether sea-level rise is currently affecting the housing market. Are prospective buyers choosing to by-pass low-elevation coastal housing, resulting in weakened demand and falling prices in these locations? As we will outline, the evidence is mixed. While it is clear that a great deal of property in coastal Florida is at risk from tidal and storm surge flooding, it is not clear that the housing market has integrated this risk either into reduced demand for housing or in reduced prices for houses susceptible to flooding.

In our 2018 survey of coastal south Florida residents (Palm and Bolsen 2020), we found that presenting maps of likely flooding had counter-intuitive effects: those who saw maps showing likely flooding in their own neighborhoods were less likely than others to believe that they were personally at risk. We also found that political party identification overwhelmingly shaped the impact of these maps: that Republicans and conservatives seeing such a map were even less likely to believe that climate change and sea-level rise are occurring and have even less concern about possible losses to their own property values. We found no regional differences within Florida to this response: residents of the Gulf Coast were as likely to view sea-level risk through a political lens as those living on the Atlantic coast. Based on our survey, we had no reason to conclude that Florida residents are particularly concerned with sea-level rise such that it affects their decisions on where to purchase a house.

Several studies have found evidence of the negative impact of sea-level rise on property values (McAlpine and Porter 2018; Bernstein et al. 2019; Ortega and Taspinar 2018). These studies are based on careful analysis of house price trends comparing areas that are relatively more or less vulnerable to flooding.

Because our survey findings showed little such concern on the part of coastal Florida homeowners, we wanted to get another perspective on how the housing market is currently integrating flood hazard. Our strategy was to triangulate on the issue by getting the viewpoint of those professionals working in the real estate market every day and who monitor the changing preferences of homebuyers: the real estate agents.

Several other factors make this study particularly relevant and timely. First, the amount of very useful information for prospective homebuyers has increased dramatically. In the summer of 2020, specific detailed information about properties for sale became available as the National Association of Realtors® added flood maps and other information about flood risk associated with sea-level rise to the web-based advertisement at realtor.com. At present, anyone who consults this information source can see the flood risk of the property, find out if flood insurance will be required by the mortgage lender, and receive an estimate of the cost of flood insurance.

The second issue that could affect the relationship between sea-level rise and housing demand on the coast is the changing cost of flood insurance. A proposal to put a new rate schedule for the National Flood Insurance Program, Risk Rating 2.0, was suggested. Such a system could result in some premium increases for owners of properties with substantial flood risk, resulting in a potential decrease in demand for coastal Florida housing.

At the same time, new construction is still proceeding at a rapid rate in coastal Florida, and the state is still a popular destination for people from other parts of the country or other parts of the world. Sales are brisk in coastal Florida, with some of the highest-priced property located in areas especially susceptible to sea-level rise.

The opinions of real estate agents can shed light on trends in the housing market. Do real estate agents see evidence that prospective homebuyers are taking elevation or flood vulnerability into account when searching for new homes? Has vulnerability to sea-level rise affected availability of home mortgages, or house price appraisals? Has the prospect of higher flood insurance premiums or the availability of new and dramatic flood information on a commonly used real estate website made a difference to sellers or to buyers?

This book describes the results of a fall 2020 survey of 680 licensed Florida Realtors, both real estate brokers and sales associates. We asked if they were witnessing any impact of elevation or flooding on house prices or price trends, their opinions about the real estate market for the next 5–10 years, their observations of home buyer behavior when flooding risk is disclosed, and whether the other key players in the real estate market, that is, mortgage lenders and appraisers, are responding to coastal flooding. Our analysis is both quantitative and qualitative, using the actual words the real estate agents shared with us to add context to the statistical trends. The results of this survey have important implications for understanding how the risks of climate change and sea-level rise are reflected in the housing market both now and in the near-term future.

References

Armal S, Porter J, Lingle B, Chu Z, Marston ML, Wing OE (2020) Assessing property level economic impacts of climate in the US: New insights and evidence from a comprehensive flood risk assessment tool. Climate 8(10):116. https://doi.org/10.3390/cli8100116

Bernstein A, Gustafson MT, Lewis R (2019) Disaster on the horizon: the price effect of sea-level rise. J Financ Econ 134(2):253–272. https://doi.org/10.1016/j.jfineco.2019.03.013

Elliott R (2021) Underwater: loss, flood insurance, and the moral economy of climate change in the United States. Columbia University Press, New York. https://doi.org/10.7312/elli19026

Fu X, Song J, Sun B, Peng ZR (2016) "Living on the edge": estimating the economic cost of sea-level rise on coastal real estate in the Tampa Bay region, Florida. Ocean Coast Manag 133: 11–17. https://doi.org/10.1016/j.ocecoaman.2016.09.009

Ghanbari M, Arabi M, Obeysekera J (2020) Chronic and acute coastal flood risks to assets and communities in Southeast Florida. J Water Resour Plan Manag 146(7):04020049. https://doi.org/10.1061/(ASCE)WR.1943-5452.0001245

Hinkel J, Church JA, Gregory JM, Lambert E, Le Cozannet G, Lowe J, McInnes KL, Nicholls RJ, van der Pol TD, van de Wal R (2019) Meeting user needs for sea-level rise information: a decision analysis perspective. Earth's Future 7(3):320–337. https://doi.org/10.1029/2018EF001071

Jackson LP, Jevrejeva S (2016) A probabilistic approach to 21st century regional sea-level projections using RCP and High-end scenarios. Glob Planet Chang 146:179–189. https://doi.org/10.1016/j.gloplacha.2016.10.006

Jevrejeva S, Frederikse T, Kopp RE, Le Cozannet G, Jackson LP, van de Wal RSW (2019) Probabilistic sea-level projections at the coast by 2100. Surv Geophys 40(6):1673–1696. https://doi.org/10.1007/s10712-019-09550-y

Kopp RE, Gilmore EA, Little CM, Lorenzo-Trueba J, Ramenzoni C, Sweet WV (2019) Usable science for managing the risks of sea-level rise. Earth's Future 7(12):1235–1269. https://doi.org/10.1029/2018EF001145

McAlpine SA, Porter JR (2018) Estimating recent local impacts of sea-level rise on current real estate losses: a housing market case study in Miami-Dade, Florida. Popul Res Policy Rev 37(6):871–895. https://doi.org/10.1007/s11113-018-9473-5

Nicholson JE, Clark K, Daraskevich G (2018) The Florida insurance market: an analysis of vulnerabilities to future hurricane losses. J Insur Regul 37(3). https://content.naic.org/sites/default/files/inline-files/JIR-ZA-37-03-EL.pdf

Nordstrom KF (1994) Beaches and dunes of human-altered coasts. Prog Phys Geogr 18(4):497–516. https://doi.org/10.1177/030913339401800402

Ortega F, Taşpınar S (2018) Rising sea-levels and sinking property values: Hurricane Sandy and New York's housing market. J Urban Econ 106:81–100. https://doi.org/10.1016/j.jue.2018.06.005

Palm R, Bolsen T (2020) Climate change and sea-level rise in South Florida. Springer Nature Switzerland AG, Cham. https://doi.org/10.1007/978-3-030-32602-9

Pattyn F, Morlighem M (2020) The uncertain future of the Antarctic ice sheet. Science 367(6484):1331–1335. https://doi.org/10.1126/science.aaz5487

Taylor ZJ (2020) The real estate risk fix: residential insurance-linked securitization in the Florida metropolis. Environ Plan A Econ Space 52(6):1131–1149. https://doi.org/10.1177/0308518X19896579

Taylor ZJ, Weinkle JL (2020) The riskscapes of re/insurance. Camb J Reg Econ Soc 13(2):405–422. https://doi.org/10.1093/cjres/rsaa015

Wdowinski S, Oliver-Cabrera T, Fiaschi S (2020) Land subsidence contribution to coastal flooding hazard in southeast Florida. Proc Int Assoc Hydrol Sci 382:207–211. https://doi.org/10.5194/piahs-382-207-2020

Woetzel J, Pinner, D, Samandari H, Engel H, Krishnan MI, Kampel C, Vesmel M (2020) Will mortgages and markets stay afloat in Florida? McKenzie Global Institute. Available at: https://www.mckinsey.com/~/media/mckinsey/business%20functions/sustainability/our%20insights/will%20mortgages%20and%20markets%20stay%20afloat%20in%20florida/mgi-will-mortgages-and-markets-stay-afloat-in-florida.pdf

Chapter 2
The Development of Vulnerable Real Estate in South Florida

Florida has been much misrepresented for good and ill. It is equally unsafe to rely on reports of the enthusiasts who portray it as an earthly Paradise, smiling in eternal sunshine, in which storms and frosts are unknown, and on whose fair and fruitful groves the blighting breath of winter never falls, as to judge it by the accounts of its enemies, who describe it as an unhealthy, barren wilderness of arid sand and pestilential swamp, the home of alligators and rattlesnakes, mosquitoes and malaria.

Iza Duffus Hardy, Oranges and alligators, Ward and Downey, London, 1887.

Florida remains a state of enchanted reality and shattered dreams.

Gary R. Mormino, 2002. Sunbelt dreams and altered states: a social and cultural history of Florida, 1950–2000. The Florida Historical Quarterly 81 (1), 3–21.

Abstract Coastal Florida continues to attract large numbers of homebuyers from other parts of the country and from other parts of the world, particularly Latin America. The contemporary Florida landscape is the result of numerous decisions about housing development, zoning and building standards made decades ago. These decisions were made with little regard for long-term environmental sustainability or notions that global climate change would make the coast more vulnerable to flooding. A 2018 survey of south Florida residents by the authors showed that most current residents anticipate little or no impact of sea-level rise on the values of their own homes. Other studies of house price trends have detected a dampening effect of vulnerability to sea-level rise on home sales. We concluded that we needed another perspective on this issue, and therefore conducted a survey of real estate agents.

Keywords Sea-level rise · Flood risk · Florida

© The Author(s), under exclusive license to Springer Nature Switzerland AG 2022

R. Palm, T. Bolsen, *Housing Market Response to Sea-Level Rise in Florida*, Coastal Research Library 37, https://doi.org/10.1007/978-3-030-88435-2_2

Real estate entrepreneurs have left an indelible mark on Florida. Indeed, the settlement pattern of Florida as we know it today can be best understood as a product of the efforts of late nineteenth century and early twentieth century American entrepreneurs who connected the state to the national railroad grid, developed land for agriculture and cities, and promoted the state for tourism and real estate investment, while making considerable fortunes for themselves. The "two Henry's", Henry Flagler and Henry Plant, are prime examples of the impact of this kind of entrepreneurship. During the 1880s and 1890s, Henry Flagler extended rail lines through Jacksonville and down the east coast of Florida to Miami, connecting the Atlantic coast of Florida to the great cities of the Northeast (Corliss 1960). The Flagler museum in Palm Beach describes him as "the person who literally invented modern Florida", in the process accumulating an estate valued at $12.5 billion in today's dollars (Flagler Museum n.d.). At about the same time, Henry Plant extended a rail line to Tampa, opening the Gulf Coast and southwest Florida to the rest of the railroad grid, as well as connecting this area with steamship service to Cuba and Key West (Johnson 1966; Turner 2005). Plant also accumulated wealth and influence, making him one of the richest and most powerful men in the South. Some 15 years after Plant's death, one of his biographers expressed it this way: "Moorish minarets of the Tampa Bay Hotel have (Fig. 2.1) stood as a monument to the memory of Plant, the builder-Plant, the seer of visions-Plant, the pioneer in the development of this vastly rich and unimaginably fertile region that today is known as the Gulf Coast section of South Florida" (Glover 1925).

Fig. 2.1 Henry Plant's Tampa Bay Hotel. (Source: Flickr Creative Commons)

Florida's wetlands were seen as a resource for both agriculture and new settlement. Governor Napoleon Broward launched the drainage of the Everglades in 1906 with the goal of converting the land into an "agricultural paradise" (Meindl et al. 2010). The next 50 years saw massive construction of canals through the wetlands, followed by flood control projects that resulted in in both subsidies for sugar cane growers but also extensive ecological damage (McCally 1999). Although the Everglades did not become a locus for diversified agricultural production, citrus production has remained an important part of the Florida economy, only surpassed by tourism as a source of revenue.

After World War I, a booming economy meant more disposable income and more leisure for large numbers of Americans, enabling tourism by the middle classes. Further, many veterans of World War I had been stationed in Florida during their service and were interested in returning. Florida promoters took advantage of this new interest (Turner 2015). Although malaria had not yet been eradicated, and home air-conditioning was still novel when it made its debut at the 1939 World's Fair, developers in the 1920s promoted Florida as a last frontier, a tropical paradise, and a land where with a small investment, land speculators could gain quick riches. The 1920s land boom turned wetlands into housing developments, or at least plans for these developments. The vast Florida project of draining wetlands, constructing canals, and creating "new" land from the dredging process created the foundation for many settlements that are now at risk from flooding and subsidence.

Florida's population was also boosted from what Mormino (2002) dubbed the "silver migration," the influx of retirees that began in the 1940s and continues through the present day. The notion of "retirement" to a new environment was also the product of twentieth century policies, including the Social Security Act of 1935 in tandem with private pension plans permitting workers to continue earning money wherever they chose to live, and housing subsidies through the FHA (Federal Housing Administration) and the VA (Veteran's Administration). Older Americans could now consider relocating to an environment where they no longer would have to deal with cold and snowy winters, and where they could purchase relatively inexpensive housing and benefit from low tax rates.

In the past decade, the population of Florida has grown from a combination of international migration and people moving in from other states, resulting in a rate of almost 950 new residents in Florida **per day** in 2019 (ISG World 2020). Presently, about two-thirds of Florida residents were born outside the state: in New York, Cuba, South America, the West Indies, Pennsylvania, New Jersey, Ohio and Puerto Rico, in that order (Gregory n.d.). This continuous population increase has fueled demand for new housing, and a subsequent large and healthy residential real estate industry: every year since 2016 more than 250,000 houses were sold in Florida, with nearly 300,000 single- family homes sold in 2019 alone (Statista research department 2021; US Census Bureau 2021).

2.1 A Legacy of Vulnerability

The modern landscape contains remnants of decisions made decades if not centuries ago. The pattern of roads, municipal boundaries, lot sizes and shapes, and changing zoning requirements create a template upon which contemporary buildings stand. In the United States, cities vary widely with respect to the eras in which they grew most quickly, resulting in very different patterns of housing stock. For example, cities that grew quickly during the nineteenth century and whose growth slowed in the twentieth century, such as Pittsburgh, Buffalo, Providence, or Cleveland, have a larger proportion of older buildings than cities that have grown more recently such as Las Vegas, Austin, or Orlando. In general, Florida has a larger percentage of recently built housing than the nation as a whole, but even here one can find communities with large numbers of dwellings constructed more than 50 years ago.

The first Florida land boom took place in the 1920s with the establishment of entire new communities such as Coral Gables, Miami Beach, Hialeah, and Hollywood (Mormino 2002; Turner 2015). Other developments of this era, including Fulford-by-the Sea and Okeelanta were more short-lived. The new communities frequently adopted exotic themes: based on "stories, designs, and motifs from the Arabian Nights, India, Egypt, and China" (Lane 2010). The architectural style of Addison Mizner, the designer of many of the great homes in Palm Beach and Boca Raton, was famously described as "Bastard-Spanish-Moorish-Romanesque-Gothic-Renaissance-Bull-Market-Damn the Expense Style" (Johnston 1953: 25; Dunlop 2019)."

Two examples of this kind of themed development were Opa-Locka and Aladdin City, both in Miami-Dade County. Opa-Locka was planned and developed by Glenn Curtiss, who had been a pioneer in aviation, inspired by the story of the Arabian Nights: Curtiss "would sometimes attribute his inspiration directly to [Douglas] Fairbanks's *Thief of Bagdad*" (Lane 2010). Some of the "Moorish" architecture survives, as for example the city hall which is listed in the National Register of Historic Places (Fig. 2.2).

Aladdin City was founded by a manufacturer of mail-order "kit homes" and promised a fully formed industrial city of 10,000 residents with streets and parks in a "Persian style." Manufactured housing that could be put together in a single day for a total price of as little as $513, including freight charges (Sears Homes of Chicagoland 2019). The city plan (Fig. 2.3) featured streets with names such as Sinbad, Sultan, Damascus and Cairo, converging on Ali Baba circle. This rapid development of new communities and new housing was stopped in part by a railroad embargo in 1925, followed by the 1926 Miami Hurricane. Only 2 years later an even more damaging hurricane occurred, followed by the Wall Street stock market crash of 1929, resulting in bankruptcies, and slowing the rampant property speculation.

A second real estate boom began in the mid-1950s through the 1970s, when about a third of the farmland in Florida was subdivided into residential building lots (Cumming 2006). Despite the previous excesses in the 1920s, the pace of development in the 1950s was "so spectacular, so searing, that it dulled the glitter of the

Fig. 2.2 Opa-Locka City Hall built in 1926. (Source: Creative Commons Attribution-Share Alike 4.0 International. https://commons.wikimedia.org/wiki/File:City_Hall_(Opa-Locka,_Florida)_1.jpg)

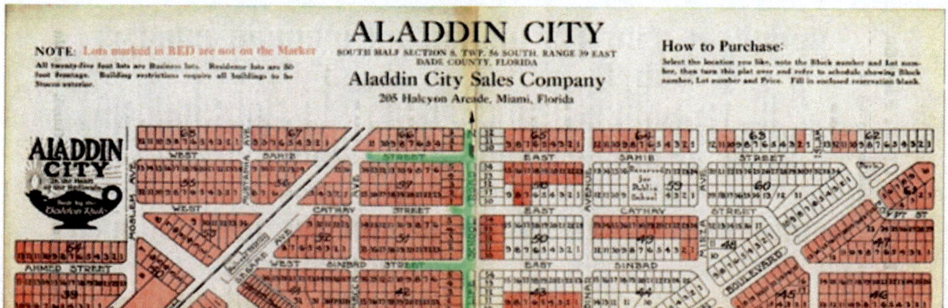

Fig. 2.3 Aladdin City, Florida. (Source: Clarke Historical Library, Central Michigan University, https://commons.wikimedia.org/wiki/File:Aladdin_City_Flordia_Land_Boom.jpg)

1920s" as investors attempted to "profit from America's greatest land rush" (Mormino 2002). Developers sold lots, some of which were actually underwater, to prospects lured to Florida for a sales pitch promising buyers a place to build their dream home for only $10.00 down and $10.00 per month for 10 years (Cumming 2006). Developments by the infamous Gulf American Corporation in Lee and Collier counties on the Gulf Coast were heavily marketed. The company founders purchased more than 100 square miles of land for $100 per acre, and dredged canals to create a "waterfront wonderland" in Lee County named Cape Coral. The developers platted the land into lots, but neglected to build water or sewer infrastructure, connections to the interstate highway, shopping centers or even grocery stores. As Grunwald (2006) colorfully described it: the company "passed off inaccessible mush as prime real estate, sold the same swampy lots to multiple buyers, and used listening devices to spy on its customers. Its hucksters spun a soggy floodplain between the Caloosahatchee River and the Gulf of Mexico as America's middle-class boomtown of the future, and suckers bought it." For the sale of lots in the company's Golden Gate Estates, the company provided "free round-trip transportation" and "free private plane rides" to out-of-state buyers unfamiliar with the area (Fig. 2.4). One of the sales points was to suggest that buyers could purchase the larger lots and later subdivide them to make even greater profits. Again, the Gulf American Land Corporation sold dreams of getting rich quickly: "This is Golden Gate Estates, where in the 1960s hastily dug canals drained the swamps, where miles of limestone roads were poured with no thought of bringing parks, shops, schools, water or sewers. Where those who did sign on the dotted line, who were in fact sold Florida swampland, came to define gullibility for a generation" (Stanley 2016).

The era in which an area is developed is not only reflected in the outline of streets and the age of its housing stock, but also in the building materials that were used and construction standards that were applied. It is notable that until 1974, the State of Florida had no statewide minimum building code. And even when the first statewide mandate was adopted, the requirement was only that local governments adopt one of four model codes to ensure minimum standards. There was no provision for statewide oversight or enforcement, and numerous local "exceptions" supported by local investment interests and construction firms were permitted. The Florida Building Code was not implemented until March 2002, 10 years after Hurricane Andrew demonstrated the deficiencies of compliance with the then-existing patchwork of local codes (Florida Housing 2017). Studies comparing houses damaged by the Hurricane Andrew with those that suffered less damage revealed that it was the newer houses that tended to be destroyed because of "low quality construction, faulty designs, and flimsy materials" (Fronstin and Holtmann 1994). In addition, despite the existence of a stringent building code in Miami-Dade and Broward counties, numerous compromises in construction practice were permitted (for example, using staples instead of nails for roof decking), as well as widespread code violations and poor enforcement procedures (Barnes and Mitrani 1992; Pita et al. 2013).

The relevance of the history of these development practices in the absence of an appropriate and widely-enforced statewide building code is clear when we consider

Fig. 2.4 Advertisement for Golden Gate Estates, Florida. (Source: Unknown author, Public domain, via Wikimedia Commons, https://commons.wikimedia.org/wiki/File:Golden_Gate_Estates,_Florida_-_Real_Estate_Sales_Brochure_Front.jpg)

the current housing stock in Florida: more than 70% of currently occupied housing – or more than 7 million houses – were built when no statewide building code was in force. Obviously, not all the substandard housing has survived, nor should we consider housing built before 2002 to be necessarily substandard. However, the development of entire towns took place in the absence of state building standards, and with little or no concern for either environmental preservation or safety from floodwater inundation, ensuring the vulnerability to sea-level rise that Florida faces today.

2.2 Do Florida Residents Perceive Their Homes as Vulnerable to the Impacts of Climate Change?

Florida's climate is one of its most important assets, and a change in this climate threatens its attractiveness to tourists, investors, and prospective homebuyers. The Environmental Protection Agency (2016) outlined some of the negative impacts of climate change for Florida including the erosion of beaches, threats to the survival of coral reefs, more frequent algae blooms such as the red tide outbreak of 2018, and an increase in the spread of insect-borne disease as well as invasive species. More hot humid summer days would endanger crop yields and threaten the health of human populations. Current projections forecast a sea-level rise of between 14 and 26 inches in the next 40 years, depending on various carbon emission reduction scenarios as well as specific locations along the Florida coast (Hauer et al. 2016; Dahl et al. 2017; Huber et al. 2017; Valle-Levison et al. 2017). The threat of coastal flooding is further enhanced by land subsidence, the settling or sinking of land, caused by, among other factors, the settling of sedimentary deposits and the withdrawal of ground water (Nicholls et al. 2021). With the combination of sea-level rise and land subsidence, Florida will see more flooding from high tides and storm surges, with cumulative resulting economic losses.

Do the thousands of people flocking to Florida worry about the vulnerability of their property to an environment susceptible to rising sea-levels and an increasing intensity of devastating storms? To answer this question, we administered a Qualtrics survey to approximately 1000 residents of coastal neighborhoods in South Florida during the last 3 months of 2018. We defined the study region, "South Florida," as that area generally south of the 28th parallel north. The metropolitan areas we sampled included seven of the 10 most populous cities in the state: Tampa-St. Petersburg-Clearwater, Cape Coral-Fort Myers, Fort Lauderdale-Pompano Beach-Deerfield Beach, Miami-Dade, West Palm Beach-Boca Raton-Delray Beach, and North Port-Sarasota-Bradenton. The full results of this survey were reported in Palm and Bolsen (2020) and are reviewed briefly here.

First, it should be noted that most of urbanized South Florida is less than 10 m (32.8 feet) above sea-level. The area is highly susceptible to even a relatively small amount of sea-level rise, particularly when the risk of flooding associated with high tides is combined with flooding from storm surge. Clearly the survey population is very much at risk from sea-level rise, at least in the next 50–100 years, if climate change continues unabated. Yet, when these coastal residents were asked whether their own home is susceptible to sea-level rise, fewer than half responded affirmatively. Respondents who identified themselves as Republicans were less likely to believe their own homes were susceptible to sea-level rise than either Independents or Democrats.

We also asked the residents about their expectations for the impacts of sea-level rise on home values. We first posed the question of whether property values **in their community** would be reduced with sea-level rise. We gave the respondents a choice of answers ranging from "not at all" to "a great deal." Overall, about one-third of the respondents said community property values would decline "a great deal", with only 28% of Republicans and 44% of Democrats giving this response. However, when asked about the impact on **their own homes**, the respondents were more optimistic with only 21% responding that their own homes would suffer "a great deal" of loss with little variation by party identification.

We showed half of the homeowners detailed maps of their local neighborhoods developed by the First Street Foundation portraying where flooding would occur in the event of a Category 3 hurricane 15 years from now, assuming continued sea-level rise. We expected that a dramatic portrayal of flood risk to their own community would elicit concern, but found, contrary to our expectations, that those who viewed the flood map were somewhat **less** likely to believe their own home is susceptible to sea-level rise and less likely to believe that the value of their own home would be negatively affected by sea-level rise. The flood map had no effect on beliefs about the extent to which sea-level rise will affect property values in their communities. Political identity did not distinguish the pattern of responses: regardless of having seen the flood map or not, and whether they identified themselves as Democrats, Independents or Republicans, most homeowners living along the coast of South Florida stated that they did not believe that sea-level rise would reduce the value of their own home.

2.3 Resolving the Conflict: A Survey of Real Estate Agents

Florida residents report that although climate change may be occurring and the sea-level may be rising, they personally are not vulnerable to the impacts of these environmental changes. The Florida Chambers of Commerce insist that South Florida remains an attractive location for real estate investment, and that one of the most attractive aspects of South Florida is its mild winter climate and its beautiful coasts (Florida Chamber of Commerce 2021). A quick perusal of real estate websites shows conclusively that new, expensive single-family houses and condominiums are being finished every day in response to a robust market. For example, in Miami Beach alone, there were more than 350 agent listings of houses for sale on Zillow priced over $3 million in early January 2021, while in the South Florida study area, there were more than 3,000 such listings, with 450 of them listed at more than $10 million. In contrast, in the United States as a whole, the median listing price for a home in December 2020 was $340,000 (Speianu and Hale 2021).

At the same time, scientists have sounded an alarm in ever-growing decibels about the vulnerability of South Florida to the effects of climate change: sea-level rise, increasing frequency and magnitude of hurricanes, and longer and hotter summers. As we will see in the following chapter, these predictions have been turned into very specific dollar estimates of the losses in Florida property value losses if reinsurers, insurers, and lenders take these predictions into account when making loan or insurance decisions. And some research has suggested this process of property devaluation in coastal Florida has already begun.

We have an unresolved conflict in opinions about what is happening in the Florida housing market. On the one hand, the 2018 survey of coastal Florida residents reported that most do not believe that if the sea-level rises, the value of their own property will be affected: only about 30% of Republican respondents and 40% of Independent or Democratic respondents said that they anticipated that that sea-level rise would affect the value of their own home. Further, many new Floridians seem to be continuing to bolster the property market, keeping demand high, particularly for properties with water access or an ocean view. On the other hand, there seems to be evidence that house prices in coastal Florida, if not falling, are not rising as fast in areas particularly susceptible to flooding. We reasoned that if property devaluation along the coast is taking place at present, it would be the real estate professionals that would be among the first to notice. Therefore, in the last 3 months of 2020 we undertook a survey of Realtors® doing business in South Florida. We again selected the southern part of coastal Florida as our study area, in part so that we could compare the responses of the real estate agents with those of the residents surveyed 2 years earlier. We divided and analyzed the responses into six housing submarket areas (Fig. 2.5) as defined by the US Department of Housing and Urban Development (Tampa-St. Petersburg-Clearwater, Cape Coral-Fort Myers, Fort Lauderdale-Pompano Beach-Deerfield Beach, Miami-Dade, West Palm Beach-Boca Raton-Delray Beach, and North Port-Sarasota-Bradenton).

Before reporting on the results of the real estate agent survey, we review the evidence that scholars have used to draw conclusions about house price trends in coastal cities generally and South Florida specifically, and how flood insurance affects the calculation of ownership costs and affects home sales prices. Following that review, we will consider the ways in which political polarization has affected the acceptance of information about climate change generally and sea-level rise specifically, as well as how visual imagery such as photos or maps can increase the impact of risk information. We review the decision of the National Association of Realtors® to include visual imagery about flood hazards in all their listings and the evolution of hazards disclosure practices within the real estate profession. We describe the various sources of information about flood vulnerability that prospective homebuyers now have, and the significance of the real estate agent in the home purchase decision. We then review the survey methodology and its findings, as well as present a detailed description of real estate agent responses in four communities that are at particular risk of inundation.

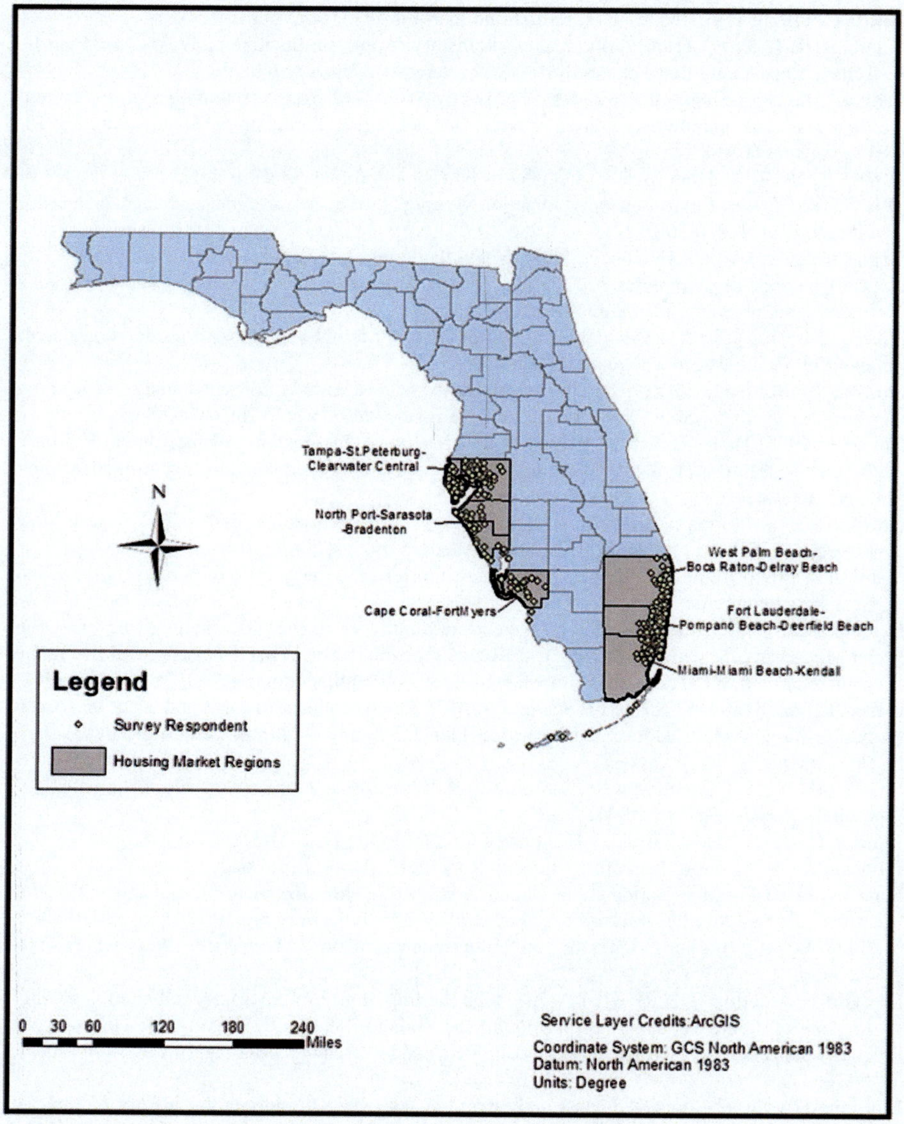

Fig. 2.5 Housing submarkets in the real estate agent survey

References

Barnes WC, Mitrani JD (1992) Code enforcement: scope and extent of problem and recommendations for solutions. Florida International University, Department of Construction Management, Technical Publication No. 105

Clarke Historical Library (2021) Aladdin City Florida land boom. Central Michigan University. Available at: https://commons.wikimedia.org/wiki/File:Aladdin_City_Flordia_Land_Boom.jpg

Corliss CJ (1960) Henry M. Flagler: railroad builder. Florida Hist Q 38(3):195–205

Cumming JB (2006) A brief Florida real estate history. Appraisal Institute of West Coast Florida, Tampa. http://mapoftheweek.net/Post/342/floridarealestatehistory.pdf

Dahl KA, Spanger-Siegfried E, Caldas A, Udvardy S (2017) Effective inundation of continental United States communities with 21st century sea-level rise. Elementa 5:37. https://doi.org/10.1525/elementa.234

Dunlop B (2019) Addison Mizner: architect of fantasy and romance. Rizzoli, New York

EPA: United States Environmental Protection Agency (2016) What climate change means for Florida. EPA 430-F-16-011

Flagler Museum (n.d.). Available at: https://www.flaglermuseum.us/history/flagler-biography

Florida Chamber of Commerce (2021) The Florida scorecard: metrics to help secure Florida's future. Available at: https://thefloridascorecard.org/

Florida Housing (2017). Available at: https://www.floridahousing.org/docs/default-source/aboutflorida/august2017/august2017/tab4.pdf

Fronstin P, Holtmann A (1994) The determinants of residential property damage caused by Hurricane Andrew. South Econ J 61(2):387–397. https://doi.org/10.2307/1059986

Glover F (1925) Henry B. Plant – genius of the west coast. Essays and Poems (Lit2Go Edition). Available at: https://etc.usf.edu/lit2go/70/florida-essays-and-poems/4604/henry-b-plant-genius-of-the-west-coast/

Gregory J (n.d.) Florida migration history 1850–2017. The great depression in Washington State Project. Available at: https://depts.washington.edu/moving1/Florida.shtml

Grunwald M (2006) The swamp: The Everglades, Florida, and the politics of paradise. Simon and Schuster, New York

Hardy ID (1887) Oranges and alligators: sketches of South Florida life. Ward and Downey, London

Hauer ME, Evans JM, Mishra DR (2016) Millions projected to be at risk from sea-level rise in the continental United States. Nat Clim Chang 67:691–695. https://doi.org/10.1038/nclimate2961

Huber JE, Van de Reit K, Sandell J, Scarpa L (2017) Salty urbanism: towards and adaptive coastal design framework to address sea-level rise. Plan J 2(2):389–414. https://doi.org/10.15274/tpj.2017.02.02.06

ISG World (2020) Miami report. Available at: https://issuu.com/isgworld/docs/isg_mr-2020-digital.1?fr=sMjFjNjYwODY2

Johnson DS (1966) Henry Bradley Plant and Florida. Florida Hist Q 45(2):118–131

Johnston A (1953) Those incredible Mizners. Rupert Hart-Davis Ltd, London

Lane C (2010) Forging Florida's sun screen: architecture, film, orientalism, and the settling of America's final frontier. Mississippi Q 63(3):373–398. https://doi.org/10.1353/mss.2010.0002

McCally D (1999) The Everglades: an environmental history. University Press of Florida, Gainesville

Meindl CF, Alderman DH, Waylen P (2010) On the importance of environmental claims-making: the role of James O Wright in promoting the drainage of Florida's Everglades in the early twentieth century. Ann Assoc Am Geogr 92(4):682–701. https://doi.org/10.1111/1467-8306.00311

Mormino GA (2002) Sunbelt dreams and altered states: a social and cultural history of Florida, 1950–2000. Florida Hist Q 81(1):3–21

Nicholls RJ, Lincke D, Hinkel J, Brown S, Vafeidis AT, Meyssignac B, Hanson SE, Merkens J-L, Fank J (2021) A global analysis of subsidence, relative sea-level change and coastal flood exposure. Nat Clim Chang 11:338–342. https://doi.org/10.1038/s41558-021-00993-z

Palm R, Bolsen T (2020) Climate change and sea-level rise in South Florida. Springer Nature Switzerland AG, Cham. https://doi.org/10.1007/978-3-030-32602-9

Pita GL, Pinelli JP, Gurley KR, Hamid S (2013) Hurricane vulnerability modeling: development and future trends. J Wind Eng Ind Aerodyn 114:96–105. https://doi.org/10.1016/j.jweia.2012.12.004

Sears Homes of Chicagoland (2019) Aladdin City: the "Town where homes will rise almost overnight." http://www.sears-homes.com/2019/09/aladdin-city-town-where-homes-will-rise.html

Speianu S, Hale D (2021) December 2020 monthly housing market trends report. Available at: https://www.realtor.com/research/december-2020-data/

Stanley G (2016) From scam to plan: Golden Gate Estates growth to get Collier County Update. Naples Daily News, April 16, 2016. Available at: https://archive.naplesnews.com/news/local/from-scam-to-plan-golden-gate-estates-growth-to-get-collier-county-update-3001ae82-0866-5197-e053-01-375956461.html/

Statista research department (2021) Single family home sales Florida. Available at: https://www.statista.com/statistics/1100147/single-family-home-sales-florida/

Turner GM (2005) Florida railroads in the 1920s. Arcadia Publishing, Mount Pleasant

Turner GM (2015) The Florida land boom of the 1920s. McFarland & Company, Inc., Publishers, Jefferson

US Census Bureau (2021) New private housing units authorized by building permits: 1-unit structures for Florida [FLBP1FH]. FRED, Federal Reserve Bank of St. Louis. Available at: https://fred.stlouisfed.org/series/FLBP1FH

Valle-Levison A, Dutton A, Martin JB (2017) Spatial and temporal variability of sea-level rise hot spots over the eastern United States. Geophys Res Lett 55(15):7876–7882. https://doi.org/10.1002/2017GL073926

Chapter 3
House Price Trends in Coastal US Cities

Living on the Gulf Coast of Florida is like living in a dream. Beautiful beaches, blue skies, and a thriving community are all aspects of life that you want to embrace.

Lend Smart Mortgage (2021) 4 key factors to consider when buying a house on the Florida gulf coast. Available at: https://www.golendsmart.com/blog/4-key-factors-to-consider-when-buying-a-house-on-the-florida-gulf-coast

It's everyone's dream... to have a house on the beach... to wake up to the sounds of the waves lapping against the shore... to see the ocean as she changes hues throughout the day, throughout the seasons, throughout the year. It's everyone's dream to have a place to go as an escape from the mad and busy world we live in... to have a special place where our children can create the most magnificent memories...

Mymove (2020) What you need to know before buying a beach house. Available at: https://www.mymove.com/home-inspiration/decoration-design-ideas/what-you-need-to-know-before-buying-a-beach-house/

Abstract Coastal location is both an amenity adding to the value of a property, and at the same time a hazard detracting from its value. Several recent studies of house price trends in flood-prone coastal portions of the United States in general, or Florida specifically, have yielded different conclusions: some have identified a price discount for coastal property that is attributed to the vulnerability to sea-level rise while others have found no such effect. Flood insurance, which has become increasingly costly in the last decade, is mandatory for homebuyers seeking mortgage financing for properties in the areas designated by FEMA of having at least a 1% annual chance of flooding. This addition to the cost of homeownership could increase prices and reduce demand for properties in these locations.

Keywords National Flood Insurance Program · Flood insurance · Hedonic price models

© The Author(s), under exclusive license to Springer Nature Switzerland AG 2022

R. Palm, T. Bolsen, *Housing Market Response to Sea-Level Rise in Florida*, Coastal Research Library 37, https://doi.org/10.1007/978-3-030-88435-2_3

3.1 The Coast as an Amenity in the Pricing of Houses

For the past century, coastal housing in Florida has been sold as a place with beautiful weather, peaceful surroundings, and the pleasant sounds of water lapping on the shore. Florida has attracted retirees as well as investors from throughout the world. It continues to be described as "one of the hottest real estate markets in the country" due to its attractiveness as a tourist destination, its theme parks, its "business-friendly climate", and most of all for its "quality of life": "an almost ideal work-life balance with great job opportunities and near-perfect weather, culture, and recreation"[1] (Rohde 2021).

But the coast is also hazardous, particularly with climate change causing more intense hurricane winds and rising seas. Can aesthetics, perceived safety or perceived risk be translated into dollar values? If so, do these aesthetics outweigh estimates of present and future risks of damage? In sum, what is the effect of a coastal location on house prices?

It is not easy to answer these questions, although there are many attempts to valorize both amenities and risks in housing prices. The classic way to identify the effects of specific characteristics of the house or its location on its final price is the calculation of "hedonic prices" (Rosen 1974). In this calculation, the goal is to estimate the contribution of various components of the house itself, such as its square footage, construction type or number of bedrooms, on its final sales price. In addition, the characteristics of the location are also included in the calculation, such as the quality of the neighborhood or accessibility to places of work or recreation. The final product is an estimate of how much the average buyer would be willing to pay for each component (Hargrave 2020). As an example, when two properties are identical except for one characteristic, one can calculate the value of that characteristic, or the marginal willingness of buyers to pay for it, from the differences in the sales prices of the two properties. Although hedonic price analysis is probably the most common way to estimate environmental impacts on house price, it can lead to inaccurate assumptions if the modeler excludes factors that are important to buyers or includes factors that are not important. For example, if homebuyers are not aware of an amenity or a hazard such that it does not enter into their valuation of the property, the hedonic model may overestimate its significance. The accuracy of hedonic models then rests at least partly on an identification of independent variables that match the factors buyers are actually considering.

Numerous empirical studies have used hedonic price calculation, or some variant of it, to estimate the value of particular amenities to the sales price of a house. A few examples are illustrative of the range of factors that have been assessed. Bond et al. (2002) found that having a lake view added more than $250,000 to the value of houses in Cleveland. Gibbons et al. (2014) estimated the value of "English nature" on housing markets using hedonic models, finding large marginal price increases associated with gardens, green space and nearby lakes or streams. They also found

[1] https://learn.roofstock.com/blog/florida-real-estate-market

differences in house prices based on the amenities of the region, with higher priced houses located in places such as the Lake District or Northumberland. Mahan et al. (2000) estimated the marginal price effect of both the size of and distance to the nearest wetland on property values in Portland, Oregon. They found that proximity to a wetland or wetlands covering larger areas added substantially to the average house value. Landry and Hindsley (2011) found that beach quality added value to shoreline houses on Tybee Island, Georgia, although cautioning that the specific value of added beach width depends at least partly on the perceptions and expectations of the homebuyers. Similarly, Catma (2020) found increased property values associated with the width of the beach at high tide in a study of property values in Hilton Head Island, South Carolina. Hamilton and Morgan (2010) found a price premium for both view and access to the beach in a study of Pensacola Beach Florida. Kim et al. (2020) found similar price premiums for houses with proximity to the beach in Jacksonville Beach, Florida. In these empirical studies, the modelers found that environmental amenities including beach location, or a water view added to the sales price of a house.

Although empirical research has found a positive price effect for both coastal location and proximity to wetlands and lakes, there are also clear costs associated with living in these environments. Houses in coastal locations or located on rising rivers or streams can flood. At that point, the ongoing costs involved in protecting the value of structures on the property could exceed the value of the water amenity. In the case of Florida, many of the residential subdivisions were created with a goal of providing property with direct access to a beach or to a boat jetty. Much of the land originated as developers drained former wetlands and built-up residential lots from the excavated sediments or fill, making it susceptible to increased flood risk as the climate changes (Bernard 1983; Grunwald 2006).

3.2 Flood Insurance as an Additional Cost in Coastal or Riverine Flood-Prone Areas

Flood insurance can act as a buffer, protecting the resident of a flood-prone area from severe financial loss. Insurance can provide peace of mind to the homeowners about future losses, but also increases the ongoing cost of homeownership. Since flood insurance is a potentially important factor in housing costs in Florida, it is important to review how this insurance operates.

Comprehensive homeowner's insurance policies were only legalized in the United States in 1950 (Hunt 1962). Before then, various hazards such as fire, theft, or personal liability were covered in separate policies, often from different companies. At present, homeowner's insurance, if not universal, is extremely common. It is usually required by mortgage lenders to protect their investment in the property but is not required if the buyer has no mortgage. The standard home insurance policy covers the cost to repair or rebuild the home if damaged or destroyed by fire, hurricane, hail, lightning or water damage from rain or snow (such as a collapsed roof), or other disasters as listed in the policy, after a deductible paid by the homeowner (Insurance Information Institute 2020). The standard policy also pays for personal belongings lost or damaged from such causes, liability against lawsuits for injuries to individuals or property damage caused by the family or pets, and additional living expenses incurred while the home is being repaired. The standard policy does not cover earthquake, flooding or landslide damage, or damage from war, neglect, nuclear hazards, rodents, or insects. For that reason, property owners are often urged to purchase insurance covering earthquake damage or termite infestations in regions that are particularly at risk from these hazards.

Flood losses had been insured separately before the advent of the comprehensive homeowners' policy. However, many insurance companies did not offer flood insurance because they viewed flooding as uninsurable, at least partly because it violates the principle of insurance that risk should be spread over a large population. In the case of flooding, only those most at risk are likely to be insured, resulting in what is known as "adverse selection." Insurers believed that premiums would have to be too high, and that even with the collection of these premiums, the companies would not have sufficient resources to cover the damage in the event of a major flood (Kunreuther 2018). The private insurance industry had thus left a serious gap in coverage, a fact made even more evident after the very large uninsured losses due to Hurricane Betsy in 1965 (Kunreuther et al. 1978). To fill this gap, the *National Flood Insurance Program* (NFIP) was established in 1968 (United States Government Public Law 90-448 1968).

The purpose of the program was both to provide flood insurance and to reduce aggregate flood risk through requirements that would result in better management of floodplains (Congressional Research Service 2020). To participate in the National Flood Insurance Program, communities had to agree to adopt flood insurance rate maps (FIRMs) outlining flood risk zones. They also had to adopt both land-use regulations and building codes to reduce future flood-related damage. In the first years of the program, despite highly subsidized premiums, relatively few homeowners at risk of flooding purchased this insurance (Knowles and Kunreuther 2014).

In 1973, the program was strengthened through the *Flood Disaster Protection Act* (United States Government 42 U.S.C. 4002 1973), mandating that federally insured lenders such as banks, credit unions, and savings-and-loans **require** flood insurance on any new loans for buildings within a Special Flood Hazard Area (SFHA), defined as an area projected to have at least a one percent chance of flooding in any year. Premium rates were to be set based on actuarial calculations. However, premiums remained relatively low, and importantly, those structures that were already in existence enjoyed a "grandfather clause" exempting them both from stricter building code standards and rate increases. The program remained in debt to the US Treasury, and pressure mounted to make the program more self-sufficient (Horn 2019).

The *Biggert-Waters Flood Insurance Reform Act of 2012* (United States Government Public Law No. 112–141 2012) was passed to increase premium rates based on risks of flood losses. The law phased out many of the subsidies of the previous program and established a council to modernize the flood zone maps. The goal was to transfer the financial responsibility for flood recovery back, at least partly, onto the homeowners in the flood zone. When FEMA started implementing the rate increases, however, fierce opposition arose. The result was the modification and weakening of the law by the next Congress in the form of the *Homeowner Flood Insurance Affordability Act of 2014* (United States Government Public Law No. 113–89 2014). The new act limited the rate of flood insurance premium increases and ensured the continued "grandfathering" of insurance premiums at below-market rates for houses constructed before 1974 (Valacer 2015).

The subsidization of property at risk for flooding continues. Modifications have been proposed to eliminate the use of "flood zones" and instead calculate risk to individual properties using actuarial practices to set risk-based rates (Congressional Research Service 2021; Montgomery and Kunreuther 2018; Peterson 2020; Shively 2017; Simpson 2017; Kane 2019). However, these changes continue to be opposed by Congressional leadership, arguing that the use of market-based rates would make this housing unaffordable (Martin 2021; Flavelle and Cochrane 2021).

In February of 2021, the First Street Foundation released a report showing growing aggregate property losses over the next 30 years due to flooding (First Street Foundation 2021). Nine of the ten counties that would suffer the greatest losses in the nation are in Florida, and eight of these are included in our study area: Broward County (Fort Lauderdale), Miami-Dade County, Charlotte County (Port Charlotte), Lee County (Fort Myers), Sarasota County, Hillsborough County (Tampa) and Palm Beach County. The report also included calculations of the differences between the premiums being paid to the NFIP and expected financial losses: where the expected loss is higher than the premium, homeowners are being subsidized by residents elsewhere. For the State of Florida, the First Street Foundation projected a need for a 480% increase in insurance rates to cover expected financial losses: "if insurance prices in Florida were increased to reflect the economic risk these homes carry in 2021 and into the future, that added cost of ownership would have a significant negative impact to the homes' value" (First Street Foundation 2021). Whether or not a rate increase of this magnitude will be politically feasible, of course, is another question.

3.3 Flood Insurance in Florida

Florida law does not mandate homeowner's or flood insurance. However, most Florida communities do participate in the Community Rating System, a voluntary program established by FEMA in 1990 that encourages floodplain management practices that exceed the minimum requirements of the NFIP. In the CRS program, communities are evaluated on a ten-point scale, and residents of communities with effective development practices are rewarded with reductions in insurance premium rates of up to 45% (FEMA 2020a, b). In addition to these FEMA-related requirements, Florida statute requires that when a homeowners' insurance policy is issued or renewed, the insurer provide a statement advising the borrower to discuss coverage with their insurance agent (State of Florida Government statutes 627.7011 2020)

The cost of flood insurance from the NFIP varies by the type of flood zone the property is located within, as defined by the FEMA risk maps (FIRMs), as well as the time of construction, and conformance to building codes in force at that time. All policies are limited to $250,000 on the dwelling and $100,000 on personal property. Additional supplemental insurance can also be purchased privately. Building coverage includes the exterior of the structure as well as carpet, appliances, and built-ins such as cabinets, while personal property insurance covers clothing, furniture, and valuables such as artwork. Flood insurance does not cover earth movement such as a landslide, even if the movement was caused by the flood, automobiles, or structures outside the home such as swimming pools or landscaping.

Insurance from the NFIP is available both within and outside of the special flood hazard zone. The NFIP offers "preferred risk policies" for properties outside the special flood hazard zone at lower rates. In addition, most standard flood insurance policies (SFIP) and preferred risk policies (PRP) require the purchase of an additional form of insurance called the "Increased Cost of Compliance" (ICC). This policy covers the holder if rebuilding the house after the flood requires increased costs of complying with current building code standards.

Flood insurance offered by the NFIP is to be priced "based on consideration of the risk involved and accepted actuarial principles" (United States Government 42 U.S.C. §4014 1968). In fact, however, properties that were built or substantially improved before December 31, 1974, are classified separately as "pre-FIRM", that is, in existence before the FIRM maps and at present remain subsidized (Congressional Research Service 2021). Other insurance rates may be "grandfathered" at lower than market premiums under certain conditions. For example, if a house was built in one flood zone and new maps defined it as in a more flood-prone zone, the owner could maintain the old rates if the building met the established codes.

As one might conclude from the description above, the rates for a standard flood insurance policy from the NFIP are highly complex (Federal Emergency Management Agency 2018, 2019). According to FEMA memorandum W-19014, effective on January 1, 2021, the annual premium for a single-family dwelling is $1.26 per $100 of coverage per year with a basement and $1.18 per $100 coverage without a basement. In 2021, residents of zones defined as "high risk areas" (for example Zone AE) paid an annual premium of $1,003 for minimum coverage ($60,000 for the building and $25,000 for property) and a maximum of $4,144 for $250,000 for the building and $100,000 for the property (Table 3.1).

Just how big a cost burden is this flood insurance? As a measure of comparison, let's compare the cost of flood insurance in the highest risk zone in coastal Florida with the cost of homeowners' associations (HOA) dues. Typically, homeowners' fees cover the maintenance of common areas such as swimming pools and landscaping, as well as utilities and property maintenance and repairs. The average resident of Miami-Dade pays a monthly HOA fee of $415, while the average in Broward County is $341 (Owers 2017). At the same time, the monthly cost of flood

Table 3.1 January 2021 FEMA flood insurance rates for Special Flood Hazard Zones

		Annual premium	Monthly premium
Minimum $60,000 on structure and $25,000 on contents			
	with basement	$1,071.00	$89.25
	without basement	$1,003.00	$83.58
Maximum $250,000 on structure and $100,000 on contents			
	with basement	$4,130.00	$344.17
	without basement	$4,410.00	$367.50

insurance for the maximum coverage of $250,000 in the highest risk zone would be an additional $344.17 per month.

To illustrate the different impact of this cost at the high and low end of the markets, let us consider two examples. In Cape Coral Florida, several houses are listed with asking prices over $3,000,000 in FEMA zone AE. The home buyer would make a down payment of about $600,000 and take a mortgage loan for the balance. At an interest rate of 3.076%, the monthly payment of principal and interest would be $10,217, property tax would be $3,372 and homeowners' insurance would be $1,106 for a total monthly payment of $14,694. The additional flood insurance cost, with the 25 percent discount for Cape Coral's community rating system, would add another $273.94 or 1.86%. In short, in this hypothetical case flood insurance amounts to less than 2% of the monthly cost of homeownership. At the other end of the price scale, another waterfront property in Cape Coral might sell for $140,000. The down payment would be $28,000, and the total monthly payment for the mortgage, tax and homeowners' insurance would be $671. The homeowner would need to get additional flood insurance for the total cost of the mortgage loan ($122,000) and would also need to pay for the minimum mandatory flood insurance for the contents of the house, for example furniture and appliances ($25,000). At 2021 rates, for a house with a basement, this would result in a payment of $1852.20 per year discounted to $1389.15 or $115.76 per month, representing a 17% increase in monthly cost. While the less than 2% increase would probably be unnoticeable to the high-end buyer, the 17% increase could be burdensome and certainly apparent to the buyer of the less expensive house. Given these relative costs, we might expect to see more response to mandatory flood insurance at the lower end of the market rather than for higher priced homes.

3.4 Does the Cost of Flood Insurance Influence Home Buyer Behavior or House Prices?

For those purchasing a house with the assistance of a home loan from a federally insured financial institution, flood insurance is mandatory if the house is in a designated zone as defined by FEMA. As noted above, the current price for this insurance through the NFIP adds a financial burden, particularly noticeable for lower income purchasers of relatively lower priced housing who need mortgage financing. This additional cost of homeownership should signal to homebuyers that a property is at risk of flooding. But has the market for houses in coastal Florida integrated this cost, resulting in decreased demand and lower prices? It is fair to say that the evidence is mixed.

Several empirical studies have found that increases in the price of flood insurance premiums directly result in a decline in house prices. In a study in Houston, Texas, Skantz and Strickland (1987) noted a 4% decrease in house prices once an area had been flooded and insurance premiums increased. Harrison et al. (2001) found that price discounts were less than the present value of future flood insurance premiums in their study of Alachua County, Florida, while Dei-Tutu and Bin (2002) found that price discounts exceeded the present value of future flood insurance costs in Pitt County, North Carolina.

To isolate the impact of flood risk on house prices changes, the researcher needs to make several decisions. The first issue involves the scale of analysis: whether prices should be tracked for an entire country, an entire state, or within the metropolitan area, for example. As Costello and Watkins (2002) have cautioned, it is important to minimize the various influences of price change to illuminate the influence of the issue being studied, in this case, new flood zoning or increases in flood insurance costs. This argues for the scale of analysis to be limited to a defined housing market where common prices for attributes of the house or location can be reliably calculated, as opposed to the study the aggregate behavior of numerous and heterogeneous markets (Jones et al. 2005; Maclennan 1977).

The second issue is the selection of a mode of analysis. Using a hedonic price model, one can calculate the value of one of the independent variables, for example, location in a flood zone. To isolate the impact of location within a flood zone on relative house price changes, one may compare the average changes for properties inside or outside the zone. The specific variables used in the hedonic equation and technical decisions such as correcting for spatial autocorrelation or including a time-variable are left to the investigator. Alternatively, one can look at the re-sale of houses over time to track price changes. If these houses are matched with respect to their characteristics except for the one under study (e.g., floodplain location), one can estimate its effect on these changes (Mueller et al. 2018).[2]

[2] Several helpful discussions of these various methods are contained in Costello and Watkins (2002) and Osland (2010).

Several studies have tracked house price changes linked to flood hazards outside the United States. Capal (2015) tested the impact of flood risk zone location on house prices in the Czech Republic and found no significant impact: "the analyzed market does not consider it too important; what is more, it seems almost indifferent." Fuerst and Warren-Myers (2019) matched property sales data with elevation data to model the impact of flooding from sea-level rise, storm surges and tides on house prices in Melbourne, Australia. They found no price discounting for either current flooding or future sea-level rise. They attributed this lack of relationship to among other things the lack of information among homebuyers about future flood risk. Beltran et al. (2019) used repeat sales to examine house price changes in England between 1995 and 2014 for houses that had experienced inland or coastal flooding. They found an immediate discount for flooded properties that disappeared within 4 years for coastal flooding. A study looking explicitly at hazard disclosure on property values on the Kapiti Coast of New Zealand (Filippova et al. 2020) found no impact of disclosure of coastal erosion on house prices and that buyers were "ignoring the future risks to their property".

Within the United States, several empirical studies such as those of St. Louis County, Missouri (Posey and Rogers 2010), and Alachua County, Florida (Harrison et al. 2001) found that location in a designated floodplain has a negative impact on property values. Netusil et al. (2019) found large discounting when examining the relationship between the footprint of a building to the 100-year flood plain in a study of repeat sales in Portland, Oregon. Zhang (2016) also found this effect in Fargo-Moorhead but identified a greater negative impact on lower priced rather than higher priced homes.

Other US studies have produced different findings. Hino and Burke (2020) examined the impact of location in a flood zone on residential property values. They assigned individual properties one of three values: within a special flood hazard area (1% floodplain), outside the SFHA or unmapped. They matched these locations with 5.65 million deed transaction records for the period of 1996–2017. They found "little evidence that information about flood risk is fully priced in property markets", although states with stronger disclosure laws and purchases involving business buyers showed some discounting. Meldrum (2016) looked at the impact of location within the NFIP defined 100-year floodplain on house prices in Boulder, Colorado over the 1995–2012 time period. The study concluded that there was an impact on the sales prices of condominiums but not on single family detached dwellings. Further, other studies have found that the amenity value of proximity to the coast outweighs the susceptibility to flooding, as in a study of Carteret County, NC where properties "subject to wave action" have higher property values (Bin and Kruse 2006).

The impact of *perceived* risk on house prices of flood-prone property has also been studied. For example, Bakkensen and Barrage (2017) combined home valuation with information on housing stock, the value that households place on living on the waterfront, and the current and future household estimation of flood risk in Rhode Island. They found that when homeowners are unduly optimistic about their safety from future flooding, they also overvalue these homes. But it is the local heterogeneity in beliefs that accounts for the overvaluation or lack of discounting of housing susceptible to climate risk.

A series of studies have estimated perceived risk by matching data on housing sales with the average county-level opinions about climate change reported in the Yale Climate Opinion Survey (Howe et al. 2015). As Bernstein et al. (2019) phrase it, "if prices in the owner-occupied housing market are indeed driven by the opinions of investors, then we expect the community's beliefs about the effects of climate change to affect the relation between sea-level rise (SLR) exposure and real estate prices." In these studies, then, the average belief in the county is ascertained not from an analysis of local survey data, but instead projected based on the population composition of the area, and the empirical findings about opinions of people with similar demographic and geographic characteristics derived from a series of national samples: "For instance, the model projects the average belief of a Hispanic/Latina woman with a bachelor's degree or higher living in Orange County, California in 2013" (Howe et al. 2015). Such estimates of average county-level beliefs in climate change are useful, particularly if it is true that demand for housing and its ultimate pricing are actually affected by the perceptions of the average current resident of a county.

Bernstein et al. (2019) using a national sample from Zillow transactions compared properties identical in characteristics except for their exposure to sea-level rise. Those within 200 feet of the coast and at an elevation of 2 meters or less were classified as exposed to sea-level rise. The authors found an average 6.6% discount for the exposed properties, with discounts ranging from 14.7% for properties that would be inundated with a one-foot increase, to 4.4% discount for those inundated only after 6 feet of sea-level rise. They also compared their data with the Yale Climate Opinion maps showing county-level measures of beliefs of residents about sea-level rise. They found a significant relationship between these average beliefs and price discounts for the owner-occupier portion of the market. Further, they found that non-owner occupiers, for example investors, were more likely to seek a sea-level rise-related discount in price. These non-owner occupiers were described as "sophisticated buyers" in comparison with the owner-occupiers. Bernstein et al. further concluded that the release of more information to all prospective buyers will strengthen the effect of sea-level rise on discounting property value for coastal property.

In a similar study, Baldauf et al. (2020) estimated the price effects for characteristics of individual properties within 50 km (31 miles) of the coast, garnered from Zillow transactions data over the period of 1997–2017. Housing characteristics were matched with geographic locations to calculate distance to the coast and estimate of the probability of an extreme flood. They also included the percentage of the county population who believed that climate change is happening based on the calculations from the Yale Climate Opinion survey (Howe et al. 2015). They found that although proximity to the coast as an amenity has a positive impact on house prices, the belief of the average county resident in climate change has a large negative impact on house prices: all else equal, a house located in an area where most residents do not believe in climate change will sell at a higher price than a house located in an area where people are concerned with climate change.

Another paper matching average climate change beliefs in a local area with housing market response focused on the impacts of beliefs on investment in additions to single family residences (Barrage and Furst 2019). The authors found that new housing construction was curtailed in areas susceptible to sea-level rise, but that this relationship was weakened when local residents were skeptical about climate change.

Murfin and Spiegel (2020) also did a national study on the impact of relative sea-level rise on property values but came up with very different findings. They analyzed almost 19 million tax records or deeds for US houses, as well as home sales between 2015 and 2017 for properties within 30 km (18.6 miles) of the coast. They estimated time-to-inundation using both historical and projected trends in sea-level rise, as well as current elevation for each property. They examined the expected predictors for property values in their hedonic models (e.g., lot size and square footage of the dwelling) as well as premiums for beachfront, more bathrooms and recency of construction. They also tested the impact of the beliefs about the existence of climate change and worry about climate change in the county as identified by the Yale Climate Opinion survey (Howe et al. 2015). Murfin and Spiegel found no impact of either elevation or worry about climate change on house prices.

There have also been studies of price response after a major storm within a local area. In a meta-analysis of 19 studies undertaken in the United States, Daniel et al. (2009) found that location in the floodplain had a negative effect on selling price, an effect that was enhanced if there had been a recent flood or increased disclosure practices. Ortega and Taspinar (2018) analyzed the housing market data from New York City over the 2003–2017 period to detect impacts of Hurricane Sandy that struck the area in 2012. They found an immediate and large drop in house prices for property damaged by the hurricane, and a smaller and gradual reduction in prices for properties that were not damaged by it. They calculated a 7.5% discount for flood-prone houses and attributed this reduction to increased awareness of flood risks on the part of residents and businesses. Another study looked at houses that were not damaged but were located close to those that were: in a study of house price changes in Florida for the period of 1980–2000, Hallstrom and Smith (2005) identified a negative impact of Hurricane Andrew in 1992 on house prices even in areas such as Lee County that were not damaged by the storm but were "near-misses".

Others have found that the post-storm price effect is transitory. Bin and Landry (2013) examined the impact of flood zone location for Pitt County, North Carolina after storm events and found that the large immediate post-storm price fall essentially disappeared after 5 or 6 years: "once relative calm is restored the price differentials decay and disappear." Atreya et al. (2013) obtained similar results in their study of Dougherty County, Georgia.

Numerous studies have focused on house price trends in Florida. Keys and Mulder (2020) analyzed house sale volume and price over the period of 2001–2018 for 771 Florida census tracts where the population centroid was within

½ mile of the coast. They classified census tracts as either "more" or "less" exposed to sea-level rise. They found that the number of home sales in more exposed levels had declined from 2013 to 2018, although sales prices were not affected. However, between 2018 and 2020, prices in more exposed areas declined by 5% from their peak. They found no evidence of changing lender behavior, however, and concluded that current subsidized prices for federal flood insurance combined with the practice of lenders selling mortgages to Fannie Mae, Freddie Mac or the FHA have resulted in lenders "mis-pricing" the risk of climate change (Keys and Mulder 2020).

In a study of Miami-Dade County, Keenan et al. (2018) found a competition among buyers for higher elevation locations that are safer from sea-level rise, a phenomenon they termed "climate gentrification." They found that house price appreciation was positively related to elevation, or safety from nuisance flooding. They calculated house price changes for 107,984 single-family dwelling properties from 1971 to 2017 and aggregated them into municipalities or unincorporated places within the county. They found that properties at 2–3 m or 3–4 m above sea-level appreciated more in value than those in the 1–2 m or 0–1 m cohorts: "the elevation of one's home in MDC [Miami-Dade County] could matter in terms of long-term price appreciation" and that "elevation was found to have a positive effect on price appreciation in 24 of the 25 jurisdictions under study." They indicated that "the findings would suggest that a consumer preference may exist in favor of higher elevation properties" and recommended further work to understand "preferences and heuristics among relevant households."

McAlpine and Porter (2018) also studied property values in Miami-Dade County, Florida analyzing sales between 2005 and 2016. They matched properties with estimated flooding associated with Category 1 and Category 3 hurricane levels, and projected inundation to the lot and nearby roads that would occur by 2032, given projected rates of sea-level rise. They combined these data with characteristics of the property including square footage of living space in the dwelling, size of the lot, and number of bedrooms to predict the sales price per square foot of dwelling space. They concluded that not only are prices being discounted because of present tidal flooding, but also that this discounting "is growing over time and is related to the amount of exposure of the lot" (McAlpine and Porter 2018). The analysis concluded that prospective buyers are becoming more aware of the flooding risk and speculate that "buyers are also worried about SLR [sea-level rise] effect on storm surges or more generally worried about the risks associated with coastal property in low-lying areas", with this awareness "seeming to serve as mechanisms to decrease or restrict property value appreciation over the study period." (McAlpine and Porter 2018).

Other empirical studies have reached different conclusions. In a study comparing house price responses to elevation in the Tampa-St. Petersburg-Clearwater (Pinellas County) and the Miami-Dade County metropolitan area, Fu and Nijman (2021) found only "a limited market response to sea-level rise": elevations of 3 feet or below in Pinellas County show a price discount, but there was no evidence of a price discount for the properties most at risk in Miami-Dade, that is those at elevations of 2 feet or less. They also explored the hypothesis that Bernstein et al. (2019) had put forward about non-owner-occupiers being more "sophisticated" and therefore responding more to sea-level rise in seeking discounts on house prices. They found instead that the two counties differed with respect to the effect of second homeowners, with those in Pinellas County gaining higher discounts than primary home dwellers but the opposite effect in Miami-Dade.

In the studies just reviewed, the researchers did directly not seek the opinions of either homebuyers or real estate agents about the factors that buyers are actually considering in making a housing choice. We suggest that the findings of these studies would be more robust if there were direct evidence that homebuyers think about relatively small differences in elevation when making their choices.

3.5 Summary

Coastal living in Florida is both an amenity and a hazard. There is evidence that some homebuyers pay a major premium for coastal locations or coastal views. On the other hand, there are also numerous empirical studies indicating that property on the coast is either declining in value or at the very least, not appreciating as much as property that is not susceptible to flooding. Pryce et al. (2011) noted that the misalignment between flood risk and house prices may be based on the way people make decisions when they are purchasing a house. The process is so filled with a "heady mix of emotion, aspiration and symbolism, the dangers of future flooding wield little influence."

Previous studies of house price trends differ in their conclusions, partly because of differences in aggregation level, methodology and the problem of unaccounted for external influences that plague the hedonic model. With respect to aggregation, studies of an entire country or of all coastal properties in the United States are at such a scale that it would be unlikely in any case that the actual values of locational characteristics can be accurately identified. The "coastal United States" is not in any sense a property "market" in which houses and their components are interchangeable, or where elements of a house or its location can be given an average price. Proximity to the salt marsh in southeast Louisiana is very different from proximity to the sea cliffs of the northern Oregon coast. Although it is possible to aggregate all coastal land and attempt to find price trends, it is not clear that this method produces much that is of use.

The various studies also differ in methodology. Putting aside the technical issues of the normalization of variables in the multiple regression or the use of controls for spatial autocorrelation, the very fact that some of the studies use independent variables that are probably never considered by homebuyers in their decision-making, or variables that are so abstruse as to be unobservable in an ordinary setting almost predetermines that the results will vary. For example, does anyone seriously believe that homebuyers distinguish between properties that are three feet above sea-level as opposed to five feet above sea-level, unless something like a disclosure about mandatory insurance calls this fact to their attention? And although it can be argued that buyers may take into consideration the general ethos of a neighborhood in making their purchase decisions, it strains credulity to argue that they show locational preferences based on the heterogeneity of opinion in the area concerning sea-level rise.

What one can conclude from the body of research reviewed is that it is undeniable that a great deal of property in Florida is at risk from flooding, particularly if climate change persists unabated. A key question is whether flood susceptibility is actually affecting house prices in the US or in Florida. Rather than drawing inferences about the marginal prices associated with coastal living from large-scale statistical analyses, we decided a more direct approach would add to our understanding. For that reason, we decided to ask this question of professionals most familiar with the home purchase process: the real estate agents. The next two chapters will further set the stage for our analysis of their responses by reviewing how messages about the risks of climate change and sea-level rise are best communicated, and the role of real estate agents in the communication process.

References

Atreya A, Ferreira S, Kriesel W (2013) Forgetting the flood? An analysis of the flood risk discount over time. Land Econ 89(4):577–596. https://doi.org/10.3368/le.89.4.577

Bakkensen LA, Barrage L (2017) Flood risk belief heterogeneity and coastal home price dynamics: going under water? (No. w23854). National Bureau of Economic Research. https://doi.org/10.3386/w23854

Baldauf M, Garlappi L, Yannelis C (2020) Does climate change affect real estate prices? Only if you believe in it. Rev Financ Stud 33(3):1256–1295. https://doi.org/10.1093/rfs/hhz073

Barrage L, Furst J (2019) Housing investment, sea-level rise, and climate change beliefs. Econ Lett 177:105–108. https://doi.org/10.1016/j.econlet.2019.01.023

Beltran A, Maddison D, Elliott R (2019) The impact of flooding on property prices: a repeat-sales approach. J Environ Econ Manag 95:62–86. https://doi.org/10.1016/j.jeem.2019.02.006

Bernard E (1983) Lies that came true: Tall tales and hard sales in Cape Coral, Florida. Anna Publishers, Ocoee

Bernstein A, Gustafson MT, Lewis R (2019) Disaster on the horizon: the price effect of sea-level rise. J Financ Econ 134(2):253–272. https://doi.org/10.1016/j.jfineco.2019.03.013

Bin O, Kruse JB (2006) Real estate market response to coastal flood hazards. Nat Hazard Rev 7 (4):137–144. https://doi.org/10.1061/(ASCE)1527-6988(2006)7:4(137)

Bin O, Landry CE (2013) Changes in implicit flood risk premiums: empirical evidence from the housing market. J Environ Econ Manag 65(3):361–376. https://doi.org/10.1016/j.jeem.2012.12.002

Bond MT, Seiler VL, Seiler MJ (2002) Residential real estate prices: a room with a view. Journal of Real Estate Research 23(1–2):129–138. https://doi.org/10.1080/10835547.2002.12091077

Capal M (2015) Flood risk as a price-setting factor in the market value of real property. Proc Econ Finance 23:658–664. https://doi.org/10.1016/S2212-5671(15)00447-5

Catma S (2020) Non-market valuation of beach quality: using spatial hedonic price modeling in Hilton Head Island, SC. Mar Policy 115:19. https://doi.org/10.1016/j.marpol.2020.103866

Congressional Research Service (2020) National Flood Insurance Program borrowing authority. Insight. Available at: https://fas.org/sgp/crs/homesec/IN10784.pdf

Congressional Research Service (2021). National flood insurance program: the current rating structure and risk rating. Updated January 25, 2021. Available at: https://fas.org/sgp/crs/homesec/R45999.pdf

Costello G, Watkins C (2002) Towards a system of local house price indices. Hous Stud 17 (6):857–873. https://doi.org/10.1080/02673030216001

Daniel VE, Florax RJ, Rietveld P (2009) Flooding risk and housing values: an economic assessment of environmental hazard. Ecol Econ 69(2):355–365. https://doi.org/10.1016/j.ecolecon.2009.08.018

Dei-Tutu VA, Bin O (2002) Flood hazards, insurance, and house prices a hedonic property price analysis. Unpublished manuscript, East Carolina University. Available at: https://citeseerx.ist.psu.edu/viewdoc/download?doi=10.1.1.200.6549&rep=rep1&type=pdf

Federal Emergency Management Agency (2018) Flood risk database (FRD) technical reference. Available at: https://www.fema.gov/sites/default/files/2020-02/Flood_Risk_Database_Technical_Reference_Feb_2018.pdf

Federal Emergency Management Agency (2019) Memorandum W-19014 from Paul Huang for write your own (WYO) principal coordinators and the National Flood Insurance Program (NFIP) direct servicing agent. Available at: https://nfipservices.floodsmart.gov/sites/default/files/w-19014.pdf

Federal Emergency Management Agency (2020a) Community Rating System (CRS). Available at: https://www.fema.gov/press-release/20210318/community-rating-system-crs-update-0

Federal Emergency Management Agency (2020b) Community status book report: Florida. Available at: https://www.fema.gov/cis/FL.html

Filippova O, Nguyen C, Noy I, Rehm M (2020) Who cares? Future sea-level rise and house prices. Land Econ 96(2):207–224. https://doi.org/10.3368/le.96.2.207

First Street Foundation (2021) The cost of climate: America's growing flood risk. Available at: https://assets.firststreet.org/uploads/2021/02/The_Cost_of_Climate_FSF20210219-1.pdf

Flavelle C, Cochrane E (2021). Chuck Schumer stalls climate overhaul of flood insurance program. The New York Times. Available at: https://www.nytimes.com/2021/03/18/climate/chuck-schumer-fema-flood-insurance.html

Fu X, Nijman J (2021) Sea-level rise, homeownership, and residential real estate markets in South Florida. Prof Geogr 73(1):63–71. https://doi.org/10.1080/00330124.2020.1818586

Fuerst F, Warren-Myers G (2019) Sea-level rise and house price capitalisation. Available at SSRN 3359289. https://doi.org/10.2139/ssrn.3359289

Gibbons S, Mourato S, Resende GM (2014) The amenity value of English nature: a hedonic price approach. Environ Resour Econ 57(2):175–196. https://doi.org/10.1007/s10640-013-9664-9

Grunwald M (2006) The swamp: The Everglades, Florida, and the politics of paradise. Simon and Schuster, New York

Hallstrom DG, Smith VK (2005) Market responses to hurricanes. J Environ Econ Manag 50 (3):541–561. https://doi.org/10.1016/j.jeem.2005.05.002

Hamilton SE, Morgan A (2010) Integrating lidar, GIS and hedonic price modeling to measure amenity values in urban beach residential property markets. Comput Environ Urban Syst 34 (2):133–141. https://doi.org/10.1016/j.compenvurbsys.2009.10.007

Hargrave M (2020) Hedonic pricing. Investopedia. Available at: https://www.investopedia.com/terms/h/hedonicpricing.asp

Harrison DT, Smersh G, Schwartz A (2001) Environmental determinants of housing prices: the impact of flood zone status. Journal of Real Estate Research 21(1–2):3–20. https://doi.org/10.1080/10835547.2001.12091045

Hino M, Burke M (2020) Does information about climate risk affect property values? (No. w26807). National Bureau of Economic Research. https://doi.org/10.3386/w26807

Horn D (2019) Introduction to the national flood insurance program (NFIP). CRS report R44593. Available at: https://crsreports.congress.gov

Howe PD, Mildenberger M, Marlon JR, Leiserowitz A (2015) Geographic variation in opinions on climate change at state and local scales in the USA. Nat Clim Chang 5(6):596–603. https://doi.org/10.1038/nclimate2583

Hunt FJ (1962) Homeowners: the first decade. Proc Casual Actu Soc 49:12–40. Available at: https://www.casact.org/sites/default/files/database/proceed_proceed62_1962.pdf

Insurance Information Institute (2020) What is covered by standard homeowners' insurance? Available at: https://www.iii.org/article/what-covered-standard-homeowners-policy

Jones C, Leishman C, Watkins C (2005) Housing market process, urban housing submarkets and planning policy. Town Plan Rev 76(2):215–233. https://doi.org/10.3828/tpr.76.2.6

Kane J (2019) Addressing Florida's rising tide of flood exposures. Insur J. https://www.insurancejournal.com/news/southeast/2019/07/24/533723.htm

Keenan JM, Hill T, Gumber A (2018) Climate gentrification: from theory to empiricism in Miami-Dade County, Florida. Environ Res Lett 13(5):054001. https://doi.org/10.1088/1748-9326/aabb32

Keys BJ, Mulder P (2020) Neglected no more: housing markets, mortgage lending, and sea-level rise (No. w27930). National Bureau of Economic Research. https://doi.org/10.3386/w27930

Kim J, Yoon S, Yang E, Thapa B (2020) Valuing recreational beaches: a spatial hedonic pricing approach. Coast Manag 48(2):118–141. https://doi.org/10.1080/08920753.2020.1732799

Knowles SG, Kunreuther HC (2014) Troubled waters: the National Flood Insurance Program in historical perspective. J Policy Hist 26(3):327–353. https://doi.org/10.1017/S0898030614000153

Kunreuther H (2018) All-hazards homeowners' insurance: challenges and opportunities. Risk Manag Insur Rev 21(1):141–155. https://doi.org/10.1111/rmir.12091

Kunreuther H, Ginsberg R, Miller L, Sagi P, Slovic P, Borkan B, Katz N (1978) Disaster insurance protection: public policy lessons. Wiley, New York

Landry CE, Hindsley P (2011) Valuing beach quality with hedonic property models. Land Econ 87 (1):92–108. https://doi.org/10.3368/le.87.1.92

Lend Smart Mortgage (2021) 4 key factors to consider when buying a house on the Florida gulf coast. Available at: https://www.golendsmart.com/blog/4-key-factors-to-consider-when-buying-a-house-on-the-florida-gulf-coast

Maclennan D (1977) Some thoughts on the nature and purpose of house price studies. Urban Stud 14(1):59–71. https://doi.org/10.1080/00420987720080061

Mahan BL, Polasky S, Adams RM (2000) Valuing urban wetlands: a property price approach. Land Econ:100–113. https://doi.org/10.2307/3147260

Martin N (2021) Why is Schumer protecting the rich from flood insurance hikes? New Republic. Available at: https://newrepublic.com/article/161798/chuck-schumer-protecting-rich-flood-insurance-hikes

McAlpine SA, Porter JR (2018) Estimating recent local impacts of sea-level rise on current real estate losses: a housing market case study in Miami-Dade, Florida. Popul Res Policy Rev 37 (6):871–895. https://doi.org/10.1007/s11113-018-9473-5

Meldrum JR (2016) Floodplain price impacts by property type in Boulder County, Colorado: condominiums versus standalone properties. Environ Resour Econ 64(4):725–750. https://doi.org/10.1007/s10640-015-9897-x

Montgomery M, Kunreuther H (2018) Pricing storm surge risks in Florida: implications for determining flood insurance premiums and evaluating mitigation measures. Risk Anal 38 (11):2275–2299. https://doi.org/10.1111/risa.13127

Mueller JM, Lima RE, Springer AE, Schiefer E (2018) Using matching methods to estimate impacts of wildfire and post wildfire flooding on house prices. Water Resour Res 54(9):6189–6201. https://doi.org/10.1029/2017WR022195

Murfin J, Spiegel M (2020) Is the risk of sea-level rise capitalized in residential real estate? Rev Financ Stud 33(3):1217–1255. https://doi.org/10.1093/rfs/hhz134

Mymove (2020) What you need to know before buying a beach house. Available at: https://www.mymove.com/home-inspiration/decoration-design-ideas/what-you-need-to-know-before-buying-a-beach-house/

Netusil R, Moeltner K, Jarrad M (2019) Floodplain designation and property sale prices in an urban watershed. Land Use Policy 88:104112. https://doi.org/10.1016/j.landusepol.2019.104112

Ortega F, Taspinar S (2018) Rising sea-levels and sinking property values: the effects of Hurricane Sandy on New York's housing market. J Urban Econ 106:81–100. https://doi.org/10.1016/j.jue.2018.06.005

Osland L (2010) An application of spatial econometrics in relation to hedonic house price modeling. Journal of Real Estate Research 32(3):289–320

Owers P (2017) HOA fees priciest in Miami-Dade, Palm Beach counties. Sun Sentinel, March 14, 2017. Available at: https://www.sun-sentinel.com/realestate/fl-bz-hoa-fees-trulia-20170314-story.html

Peter G. Peterson Foundation (2020) Budget basics: the national flood insurance program. Available at: https://www.pgpf.org/budget-basics/the-national-flood-insurance-program

Posey J, Rogers WH (2010) The impact of special flood hazard area designation on residential property values. Public Works Manag Policy 15(2):81–90. https://doi.org/10.1177/1087724X10380275

Pryce G, Chen Y, Galster G (2011) The impact of floods on house prices: an imperfect information approach with myopia and amnesia. Hous Stud 26(02):259–279. https://doi.org/10.1080/02673037.2011.542086

Rohde J (2021) Why investors still love Florida's real estate market in 2021. Roofstock Blog. Available at: https://learn.roofstock.com/blog/florida-realestate-market. Accessed 3 Mar 2021

Rosen S (1974) Hedonic prices and implicit markets: product differentiation in pure competition. J Polit Econ 82(1):34–55. https://doi.org/10.1086/260169

Shively D (2017) Flood risk management in the USA: implications of National Flood Insurance Program changes for social justice. Reg Environ Chang 17:1663–1672. https://doi.org/10.1007/s10113-017-1127-3

Simpson AG (2017) Why federal flood program is sinking deeper into debt. Insur J. https://www.insurancejournal.com/news/national/2017/09/05/463227.htm

Skantz T, Strickland T (1987) House prices and a flood event: an empirical investigation of market efficiency. Journal of Real Estate Research 2(2):75–83. https://doi.org/10.1080/10835547.1987.12090534

State of Florida Government (2020) Statutes 627.7011 4a and 4b. https://www.flsenate.gov/laws/statutes/2020/627.7011

United States Government (1968) National Flood Insurance Program: public law 90-448. https://www.govinfo.gov/content/pkg/STATUTE-82/pdf/STATUTE-82-Pg476.pdf

United States Government (1973) Flood Disaster Protection Act:42 U.S.C. 4002. https://www.fema.gov/sites/default/files/2020-07/national-flood-insurance-act-1968.pdf

United States Government (2012) Biggert-Waters Flood Insurance Act of 2012: public law 112-141. https://www.govinfo.gov/content/pkg/PLAW-112publ141/pdf/PLAW-112publ141.pdf

United States Government (2014) Homeowner Flood Insurance Affordability Act of 2014: public law 113-89. https://www.congress.gov/113/plaws/publ89/PLAW-113publ89.pdf

Valacer J (2015) Thicker than water: America's addiction to cheap flood insurance. Pace L Rev 35:1050. Available at: https://digitalcommons.pace.edu/plr/vol35/iss3/7

Zhang L (2016) Flood hazards impact on neighborhood house prices: a spatial quantile regression analysis. Reg Sci Urban Econ 60:12–19. https://doi.org/10.1016/j.regsciurbeco.2016.06.005

Chapter 4
Effective Communication About Climate Change and Sea-Level Rise

Florida is drowning. Condos are still being built.
Can't humans see the writing on the wall?
Megan Mayhew Bergman (2019), https://www.theguardian.com/environment/2019/feb/15/florida-climate-change-coastal-real-estate-rising-seas

Abstract In the United States, people are deeply divided over the issue of climate change. Ideological and partisan divisions have been found to have a major influence on the way homebuyers evaluate flood risks to their homes and their neighborhoods. The survey reported in this book focused on the views of real estate agents, and their understanding of the way in which prospective homebuyers perceive flood risk related to sea-level rise. To put these responses into context, this chapter synthesizes some of the social science research about factors that affect risk perception, as well as the effectiveness of communication strategies that use salient, local and visual imagery.

Keywords Partisanship · Public opinion · Message framing · Polarization · Risk communication

The key question we are investigating is whether the housing market has integrated the risk of climate change, and specifically flooding associated with sea-level rise, into house prices. We wish to know whether homebuyers think about the increasing hazard of flood risks associated with sea-level rise when they search for homes, avoid properties in low-lying coastal locations, and in the process, lower both demand and ultimately sales prices. If this is occurring, we posit that real estate agents would have observed buyers seeking information about the elevation of the property, requirements for flood insurance, and records of past flooding as well as prospects for future flooding given sea-level rise.

In this chapter, we review previous research that addresses what Americans believe about climate change and sea-level rise and how these concepts have become politicized. We conclude with a discussion of how we expect real estate agents to respond to the provision of specific information about the flood risk to individual

© The Author(s), under exclusive license to Springer Nature Switzerland AG 2022

R. Palm, T. Bolsen, *Housing Market Response to Sea-Level Rise in Florida*, Coastal Research Library 37, https://doi.org/10.1007/978-3-030-88435-2_4

properties that is now mandated by their professional trade association for their own listings and for those they sell.

4.1 Beliefs About Climate Change and Sea-Level Rise

The factors that influence and shape Americans' beliefs about climate change have been well-documented over the past 20 years. Although there has been a near-universal consensus among climate scientists that human activity is the dominant cause of accelerated global warming (Cook et al. 2016), many Americans do not believe the consensus and express doubt about the science (Leiserowitz et al. 2018; Egan and Mullins 2017). This doubt has been observed to be more common in Republicans than in Democrats, with partisan differences in opinion as to whether human activity is contributing a great deal to climate change, and that climate change is impacting their local community (Tyson and Kennedy 2020). Partisans are also deeply divided on whether they will have to move to another location because of the effects of climate change (Kim et al. 2021), and over the extent to which they support government action, such as a tax on carbon, to address climate change (Dunlap et al. 2016; Feldman et al. 2012; Hart et al. 2015; Kerr 2020; Bord et al. 2000; Hornsey et al. 2016; Guber 2013; McCright and Dunlap 2011; Egan and Mullin 2017).

One of the products of climate change is global sea-level rise (Golledge 2020; Shukla et al. 2017). Global warming causes sea-level rise because of an increase in the amount of water in the oceans from melting ice sheets and glaciers, and the expansion of the water itself as seas get warmer (Frederikse et al. 2020). A recent study showed that 110 million people in the world currently live below the high tide line, and 230 million live less than one meter above sea-level (Kulp and Strauss 2019). Sea-level rise, like climate change, is viewed through a political lens in the United States with Republicans less likely than Democrats to acknowledge its existence (Palm and Bolsen 2020; Covi and Kain 2016). In addition, the effects of sea-level rise, such as flooding or coastal erosion, are often not directly associated with climate change, and are seen as less salient (Akerlof et al. 2017).

In addition to political partisanship, other factors affect attitudes about climate change, and particularly beliefs about vulnerability to its negative effects. Most generally, psychologists have posited that people who are at risk may try to resolve the conflict between what they know to be true, and what they want to be true. This resolution of the conflict between actual risk and perceived vulnerability, may express itself in statements such as "the hazard of flooding is exaggerated and won't affect me," or "my house won't be damaged even if others in my neighborhood will."

The association of other individual characteristics with climate change beliefs has also been explored in survey research. Some surveys have found that women and persons of color tend to express more belief in climate change and concern about its consequences (Finucane et al. 2000; Leiserowitz 2005, 2006; McCright and Dunlap 2011; Schuldt and Pearson 2016). Some have found that higher levels of education are associated with greater concern about climate change (Hornsey et al. 2016; van der Linden et al. 2015), but others have found that more education simply increases belief polarization (Bolsen et al. 2015; Malka et al. 2009). There are also mixed findings regarding any association between income or degree of religiosity and beliefs about climate change (Hornsey et al. 2016; Smith and Leiserowitz 2012).

Age is also an inconsistent predictor of beliefs: while some studies find that older people are more likely to believe the climate change is natural and not affected by human behavior (Heath and Gifford 2006; Hornsey et al. 2016; Kellstedt et al. 2008; Malka et al. 2009; Milfont 2012), other research calls into question the robustness of this relationship (Akerlof et al. 2016; O'Conner et al. 1999; Sundblad et al. 2007; Slimak and Dietz 2006). Regardless of age, prior experience with living in particularly vulnerable areas may also influence beliefs about climate change (Akerlof et al. 2015; Capstick and Pidgeon 2014; Howe et al. 2014; Krosnick et al. 2006; Reser et al. 2014; van der Linden et al. 2015; Zia and Todd 2010).

Numerous studies have linked people's "accessible experiences" with local weather and temperature with their beliefs about climate change (Egan and Mullin 2012; Hamilton and Stampone 2013; Joireman et al. 2010; Shao et al. 2014; Schuldt and Roh 2014; Weber and Stern 2011; Zaval et al. 2014). It is also common that soon after the disaster occurs, many people forget its lessons, showing what can be called a "collective amnesia" (Kousky et al. 2020; Marshall 2015).

There are consistent findings that people's values and worldviews also have an independent effect on their beliefs about climate change (Kahan 2013). Cultural cognition theory, for instance, explains why individuals who possess values oriented towards "individualism" and "hierarchy" generally express less belief in climate change and support for actions to address the problem compared to people who value "egalitarianism" and "communitarianism" (Kahan et al. 2011; Kahan 2015). Moreover, as people with different values become more "scientifically literate" – with higher levels of numeracy and cognitive reflection, for instance – they become more adept at distinguishing between "scientific conclusions" that support or contradict their own values and cultural commitments when evaluating new information and updating their beliefs (Kahan et al. 2012).

Real estate agents, of course, are affected by many of the same factors as homebuyers when considering the risk of sea-level rise: their political orientation, previous experience with the local area, gender, scientific literacy, and cultural worldview. And just as homebuyers may dismiss information about risks of climate change because they conflict with previously held ideas, so too may real estate agents. But there is an additional factor at work for real estate agents. Since their primary goal is to sell property, their economic interests may add even more motivation to take a very optimistic view of the safety of the property and minimize its potential for environmental deterioration. As they are looking out for "the customer's best interest when in the process of making a sale" (Dunlap et al. 1988), they are also looking out for the long-term prospects of their own business interests. Or, as Upton Sinclair put it in 1934, "It is difficult to get a man to understand something, when his salary depends on his not understanding it" (Sinclair 1994). Social scientists describe this process of evaluating information with specific goals or outcomes in mind as "motivated reasoning." In this instance, people evaluate information in pursuit of a "directional goal," which in this case is to maintain the fiction that the property they are selling or investing in is not at risk (Bolsen et al. 2014; Druckman 2013; Kunda 1990; Kraft, Lodge and Taber 2015). In short, the real estate agent may have another, more self-interested reason to reject the science around climate change and sea-level rise, or even to deny trends that are occurring if they run counter to their business interests.

But, to repeat, for most people the driving factor affecting climate change beliefs in the United States is political ideology. How and why have partisan politics become so important in the assessment of risk associated with climate change?

4.2 How the Concept of Climate Change Became Polarized

Republicans and Democrats were not always so divided on the issue of climate change. In 1988, when the United Nations created the Intergovernmental Panel on Climate Change (IPCC), political conservatives expressed the same level of concern about "global warming" as liberals (Bolsen and Shapiro 2018). However, by the late 1990s, partisan divisions on climate change had begun to emerge (Nisbet and Myers 2007). These divisions have grown partly as the result of advocacy groups that disputed climate science (Oreskes and Conway 2010; McCright and Dunlap 2000; Jacques et al. 2008; Hmielowski et al. 2014). The debate moved into the political realm, and the voices of contrarians were strengthened as journalists attempted to offer audiences "balanced coverage" (Boykoff and Boykoff 2004, 2007). Cable news and partisan media outlets added to the polarization, as is demonstrated by a contrast in coverage of climate change by CNN or MSNBC as opposed to Fox news (Feldman et al. 2012; Hmielowski et al. 2014; Gauchet 2012; Gustafson et al. 2019). The statements and positions taken by Republican leaders as well as conservative media commentators have also served as a powerful polarizing force (Brulle et al. 2012; McCright and Dunlap 2003; Tesler 2018). For example, in response to the release of the 4th National Climate Assessment Report in November 2018, President Trump stated, "I don't believe it," and suggested its conclusions were based on the most extreme scenarios and "fake science" promoted by the left-wing media to scare people into supporting a radical "political agenda" (Lavelle 2018).

The result of this politicization of climate change has been an increased resistance by at least some Americans to new scientific information. Instead, people may tend to seek out information that confirms their existing beliefs and avoid or disregard new information that contradicts their views (Kunda 1990; Druckman 2012; Druckman and Bolsen 2011; Lodge and Taber 2013; Ma et al. 2019).

This resistance to new information is affected by several factors. First, people may want to protect their group-based identities: Republicans may reject a message about the importance of action on climate change, for instance, to conform to their perceived political in-group's position, whereas Democrats might be motivated to accept the message to bolster their own group's views (Mildenberger and Tingley

2019; Tesler 2018; Unsworth and Fielding 2014). Second, people may also dismiss information about the risks of sea-level rise because it threatens their core values. For example, people who value free markets may reject the scientific fundamentals of climate change because they may suggest the need for more government regulation on land use and/or restrictions on personal liberties (Lewandowsky et al. 2013; Campbell and Kay 2014). Third, the new information may be presented at a time when commitment to a decision has already taken place: in the case of disclosure of flood risk associated with the purchase of a home, the information may be integrated as a mere distraction (Stern 2005). Fourth, the rise of digital media and the transformation of media platforms has also given rise to unique challenges given the increased ease with which misinformation about climate change can proliferate (Lazer et al. 2018; Iyengar and Massey 2019; Osmundsen et al. 2020; Levy et al. 2021). For all these reasons, efforts to provide more information to skeptical audiences can sometimes "backfire," causing even more entrenchment in skeptical beliefs (Byrne and Hart 2009; Hart and Nisbet 2012; Nisbet et al. 2015; Nyhan and Reifler 2010; Palm et al. 2020; Zhou 2016).

4.3 Using Visual Imagery to Convey Risks of Sea-Level Rise

Despite this resistance to new information about climate change and its relationship to sea-level rise, several studies have identified communication strategies to reach even these skeptical audiences (Bolsen and Shapiro 2017; Druckman and Lupia 2017; Goldberg et al. 2019). Maps, moving pictures and photos have been shown to have powerful impacts on public opinion (Hart and Feldman 2016; Leiserowitz 2006; O'Neill 2013; Sheppard et al. 2011). These images are particularly effective when they reduce the "psychological distance" between the viewer and the flood risk. For example, images have a greater impact if they portray the local effects of sea-level rise and climate change on audiences as opposed to its effects on exotic animals or locations such as polar bears at the North Pole (Bolsen et al. 2018, 2019; Feldman and Hart 2018; Lorenzoni et al. 2007; Myers et al. 2012; Pidgeon and Fischoff 2011; Retchless 2018). An innovative example is the King Tides Project (described at kingtides.net), which began in Australia in 2009 and is now underway in the US and Canada. It capitalizes on the power of visual imagery to influence public opinion by encouraging coastal residents to take pictures of flooding at high tide and post the images online (Akerlof et al. 2017). By encouraging this kind of active engagement with tidal flooding, the project increases the perceived importance of sea-level rise as a local and imminent threat.

Another demonstration of the effects of using visual imagery to increase awareness and change attitudes was contained in a survey-experiment by Bolsen et al. (2018) that evaluated the effects of text and visual imagery highlighting the effects of rising sea-levels on Boston and Miami. They found that an animated map showing the areas of these cities that would be flooded under conditions of climate change resulted in an increased belief that climate change will cause negative results, an increased concern for coastal communities and even increased belief that global warming is occurring. Moreover, these frames were impactful on Republicans, Democrats, and Independents. When the researchers presented this imagery in competition with climate science politicization (Bolsen et al. 2019), they found that although the presence of politicization reduced the otherwise positive impact of textual messages, animated visual imagery illustrating the projected effects of climate change bolstered the relative impact of scientific messages. These studies and others show that visual frames work to combat the effects of politicization (van der Linden et al. 2014).

The use of images as a way of conveying risk is not always effective, however. Well-intentioned presentations of images can backfire due to varying motivations that drive how local audiences process otherwise engaging visual imagery. For instance, Mildenberger et al. (2019) studied how providing residents in the San Francisco Bay Area with personally relevant risk messages influenced climate risk perceptions in the context of sea-level rise. The authors showed respondents one of two maps that depicted sea-level rise flooding coupled with a 100-year storm scenario. They found that exposure to a map showing flooding at the zip-code level significantly *decreased* respondents' belief that sea-level rise would threaten the future of their communities. Our south Florida survey in 2018 had similar findings in that the presentation of a local map decreased concern with the vulnerability of the respondents' own homes (Palm and Bolsen 2020). One explanation of such "backfire effects" is that respondents may be motivated to protect their prior belief that their property was and will remain a good investment, and therefore simply reject the new information.

From these studies and others like them, we can conclude that messages that highlight the local effects of climate change clearly can break through belief polarization and politicization. Moser (2013) found that strong emotional attachment to place transcends other issues when engaging people to come up with solutions to sea-level rise or climate change. Her research and that of others suggest that emphasizing the geographically proximate and personal impacts of climate change may help promote awareness and the adoption of mitigation measures (Nisbet 2014; Leiserowitz 2007; Morton et al. 2011; Spence and Pidgeon 2010). In sum, based on these empirical findings we would expect that imagery, including maps, motion-pictures, and photos, would have an impact on the understanding of climate change risk, particularly if these images are created of the local area. And just such images were created and promulgated in 2020.

4.4 The Addition of Visual Imagery About Flood Hazards to realtor.com

Listings sold by members of the National Association of Realtors® are compiled on the Multiple Listing Service and listed on the website realtor.com.[1] In August of 2020, the National Association of Realtors® integrated the flood classifications derived by the First Street Foundation[2], termed "flood factor" into all of its property listings. In an article reassuring Realtors® of the utility of this new feature (Tracey 2020), the National Association of Realtors® noted that members could avoid liability in seller disclosures and "better manage client expectations" as well as keeping themselves "at the center of the transaction and as the primary source of property information."

Before the addition of this information, the typical real estate listing contained information on the size, age, and price of the dwelling, along with features that the seller or real estate agent wanted to highlight to prospective buyers such as fencing, built-in appliances and convenient location. After August 26, 2020, all listings added a box indicating such information as commute time and noise level. They also provided the FEMA Zone and a numerical score for the Flood Factor (from 1 to 10). By clicking on this map, the viewer is taken to another map neighborhood map where they can see the location and rating of local schools, crime rates, "lifestyle" (e.g., the location of grocery stores, restaurants), transit (current traffic, public transit, and bike lanes), and "flood." The flood map portrays the depth of flood waters associated with the projected flood-risk for the property and neighborhood over a 30-year period (Fig. 4.1).

Prior research on the addition of local visual imagery suggests that these new maps should have had some impact on prospective buyers. If so, the implications for local real estate brokers could be serious. Given the mixed motivations of real estate agents, to provide information to clients, to avoid lawsuits resulting from non-disclosure, and to maximize the sales price of any property they are working with, we would expect a strong reaction to this change. In the following chapters, we will report the response of real estate agents to the addition of this imagery, and their views on the impacts this information is having on both buyers and sellers.

[1] The company that operates this website is Move Company, which in turn is owned by News Corporation, a media and publishing company founded by Rupert Murdoch that also owns Dow Jones & Company, the Wall Street Journal, and HarperCollins book publishers.

[2] The First Street Foundation is a 501(3)(c) foundation headquartered in Brooklyn, NY with a stated mission of "defining America's flood risk."

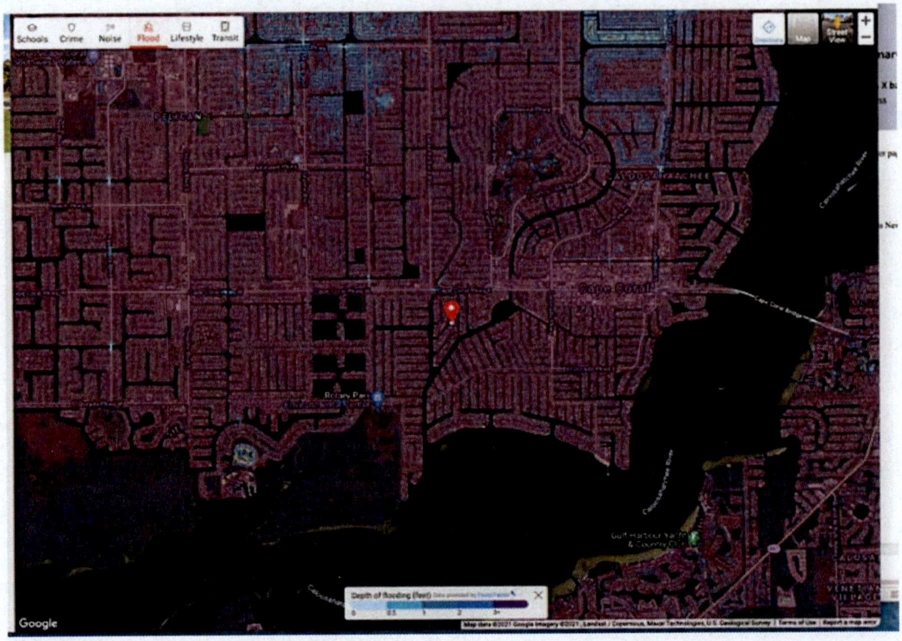

Fig. 4.1 Projected flooding in Cape Coral, Florida. Area colored in purple would have 3 or more feet of floodwater. (Source: Flood risk data is provided by Flood Factor®, a product of First Street Foundation®)

4.5 Summary

A large body of social science research informs the question of how homebuyers might integrate information about climate change and sea-level rise in their search for a new home in Florida. Numerous surveys have found that views about the reality of climate are affected by political party affiliation in the United States, and that Republicans tend to be more skeptical about climate change than Democrats. Other factors affecting general beliefs have also been explored, including prior experience with flooding and general worldview. The politicization of the concept of climate change has a relatively short history but is nonetheless a very powerful force in affecting risk perception and subsequent behavior. Since real estate agents are not immune to the general worldviews shared by people in their home communities, they too are likely to share the same opinions as those in their market area. In addition, they may be motivated by their business interests to minimize information about the impacts of rising sea-levels in their market areas.

Despite a resistance to new information about the reality of climate change and sea-level rise on the part of some Americans, there is also evidence that certain forms of communication can be effective with skeptical audiences, particularly communication that is local, salient, and visual. In the next chapters we will explore the response of real estate agents to the flood hazard disclosure that is now part of the

National Association of Realtors® website listings, along with their general opinions about whether homebuyers in Florida are avoiding low-lying property, resulting in reduced demand and ultimately lower prices. Finally, we will review their expectations concerning the impact of flooding on housing demand and sales prices in coastal Florida over the next 5–10 years.

References

Akerlof K, Delamater P, Boules C, Upperman C, Mitchell C (2015) Vulnerable populations perceive their health as at risk from climate change. IJERPH 12:15419–15430. https://doi.org/10.3390/ijerph121214994

Akerlof KL, Rowan KE, La Porte T, Batten BK, Ernst H, Sklarew DM (2016) Risky business: engaging the public on sea-level rise and inundation. Environ Sci Pol 66:314–323. https://doi.org/10.1016/j.envsci.2016.07.002

Akerlof K, Covi M, Rohring E (2017) Communicating sea-level rise. In: Oxford research encyclopedia of climate science. Oxford University Press, Oxford. https://doi.org/10.1093/acrefore/9780190228620.013.417

Bergman MM (2019) Florida is drowning. Condos are still being built. Can't humans see the writing on the wall? The Guardian. Available at: https://www.theguardian.com/environment/2019/feb/15/florida-climate-change-coastal-realestate-rising-seas

Bolsen T, Shapiro MA (2017) Strategic framing and persuasive messaging to influence climate change perceptions and decisions. In: Oxford research encyclopedia of climate science. Oxford University Press, Oxford. https://doi.org/10.1093/acrefore/9780190228620.013.385

Bolsen T, Shapiro MA (2018) The US news media, polarization on climate change, and pathways to effective communication. Environ Commun 12:149–163. https://doi.org/10.1080/17524032.2017.1397039

Bolsen T, Druckman JN, Cook FL (2014) The influence of partisan motivated reasoning on public opinion. Polit Behav 36:235–262. https://doi.org/10.1007/s11109-013-9238-0

Bolsen T, Druckman JN, Cook FL (2015) Citizens', scientists', and policy advisors' beliefs about global warming. Ann Am Acad Polit Soc Sci 658:271–295. https://doi.org/10.1177/0002716214558393

Bolsen T, Kingsland J, Palm R (2018) The impact of frames highlighting coastal flooding in the USA on climate change beliefs. Clim Chang 147:359–368. https://doi.org/10.1007/s10584-018-2143-0

Bolsen T, Palm R, Kingsland JT (2019) Counteracting climate science politicization with effective frames and imagery. Sci Commun 41(2):147–171. https://doi.org/10.1177/1075547019834565

Bord RJ, O'Connor RE, Fisher A (2000) In what sense does the public need to understand global climate change? Public Understand Sci 9:205–218. https://doi.org/10.1088/0963-6625/9/3/301

Boykoff MT, Boykoff JM (2004) Balance as bias: global warming and the US prestige press. Glob Environ Chang 14:125–136. https://doi.org/10.1016/j.gloenvcha.2003.10.001

Boykoff MT, Boykoff JM (2007) Climate change and journalistic norms: a case-study of US mass-media coverage. Geoforum 38:1190–1204. https://doi.org/10.1016/j.geoforum.2007.01.008

Brulle RJ, Carmichael J, Jenkins JC (2012) Shifting public opinion on climate change: an empirical assessment of factors influencing concern over climate change in the U.S., 2002–2010. Clim Chang 114:169–188. https://doi.org/10.1007/s10584-012-0403-y

Byrne S, Hart PS (2009) The boomerang effect: a synthesis of findings and a preliminary theoretical framework. Ann Int Commun Assoc 33:3–37. https://doi.org/10.1080/23808985.2009.11679083

Campbell TH, Kay AC (2014) Solution aversion: on the relation between ideology and motivated disbelief. J Pers Soc Psychol 107:809–824. https://doi.org/10.1037/a0037963

Capstick SB, Pidgeon NF (2014) Public perception of cold weather events as evidence for and against climate change. Clim Chang 122:695–708. https://doi.org/10.1007/s10584-013-1003-1

Cook J, Oreskes N, Doran PT, Anderegg WRL, Verheggen B, Maibach EW, Carlton JS, Lewandowsky S, Skuce AG, Green SA, Nuccitelli D, Jacobs P, Richardson M, Winkler B, Painting R, Rice K (2016) Consensus on consensus: a synthesis of consensus estimates on human-caused global warming. Environ Res Lett 11:048002. https://doi.org/10.1088/1748-9326/11/4/048002

Covi MP, Kain DJ (2016) Sea-level rise risk communication: public understanding, risk perception, and attitudes about information. Environ Commun 10:612–633. https://doi.org/10.1080/17524032.2015.1056541

Druckman JN (2012) The politics of motivation. Crit Rev 24:199–216. https://doi.org/10.1080/08913811.2012.711022

Druckman JN (2013) Stunted policy support. Nat Clim Chang 3:617. https://doi.org/10.1038/nclimate1939

Druckman JN, Bolsen T (2011) Framing, motivated reasoning, and opinions about emergent technologies. J Commun 61:659–688. https://doi.org/10.1111/j.1460-2466.2011.01562.x

Druckman JN, Lupia A (2017) Using frames to make scientific communication more effective. In: Oxford handbook of the science of science communication. Oxford University Press, Oxford, pp 351–360. https://doi.org/10.1093/oxfordhb/9780190497620.013.38

Dunlap BJ, Dotson MJ, Chambers TM (1988) Perceptions of real-estate brokers and buyers: a sales-orientation, customer-orientation approach. J Bus Res 17:175–187. https://doi.org/10.1016/0148-2963(88)90050-1

Dunlap RE, McCright AM, Yarosh JH (2016) The political divide on climate change: partisan polarization widens in the U.S. environment. Sci Policy Sustain Dev 58:2–4. https://doi.org/10.1080/00139157.2016.1208995

Egan PJ, Mullin M (2012) Turning personal experience into political attitudes: the effect of local weather on Americans' perceptions about global warming. J Polit 74:796–809. https://doi.org/10.1017/S0022381612000448

Egan PJ, Mullin M (2017) Climate change: US public opinion. Annu Rev Polit Sci 20:209–227. https://doi.org/10.1146/annurev-polisci-051215-022857

Feldman L, Hart PS (2018) Is there any hope? How climate change news imagery and text influence audience emotions and support for climate mitigation policies. Risk Anal 38:585–602. https://doi.org/10.1111/risa.12868

Feldman L, Maibach EW, Roser-Renouf C, Leiserowitz A (2012) Climate on cable: the nature and impact of global warming coverage on Fox News, CNN, and MSNBC. Int J Press/Polit 17:3. https://doi.org/10.1177/1940161211425410

Finucane ML, Slovic P, Mertz CK, Flynn J, Satterfield TA (2000) Gender, race, and perceived risk: the "white male" effect. Health Risk Soc 2:159–170. https://doi.org/10.1080/02732170490459485

Frederikse T, Landerer F, Caron L, Adhikari S, Parkes D, Humphrey VW, Dangendorf S, Hogarth P, Zanna L, Cheng L, Wu Y-H (2020) The causes of sea-level rise since 1900. Nature 584:393–397. https://doi.org/10.5281/zenodo.3862995

Gauchat G (2012) Politicization of science in the public sphere: a study of public trust in the United States, 1974 to 2010. Am Sociol Rev 77:167–187. https://doi.org/10.1177/0003122412438225

Goldberg MH, van der Linden S, Ballew MT, Rosenthal SA, Gustafson A, Leiserowitz A (2019) The experience of consensus: video as an effective medium to communicate scientific agreement on climate change. Sci Commun 41(5):659–673. https://doi.org/10.1177/1075547019874361

Golledge NR (2020) Long-term projections of sea-level rise from ice sheets. WIREs Clim Change 11(2). https://doi.org/10.1002/wcc.634

Guber DL (2013) A cooling climate for change? Party polarization and the politics of global warming. Am Behav Sci 57(1):93–115. https://doi.org/10.1177/0002764212463361

Gustafson A, Rosenthal SA, Ballew MT, Goldberg MH, Bergquist P, Kotcher JE, Maibach EW, Leiserowitz A (2019) The development of partisan polarization over the Green New Deal. Nat Clim Chang 9:940–944. https://doi.org/10.1038/s41558-019-0621-7

Hamilton LC, Stampone MD (2013) Blowin' in the wind: short-term weather and belief in anthropogenic climate change. Weather Clim Soc 5:112–119. https://doi.org/10.1175/WCAS-D-12-00048.1

Hart PS, Feldman L (2016) The impact of climate change–related imagery and text on public opinion and behavior change. Sci Commun 38:415–441. https://doi.org/10.1177/1075547016655357

Hart PS, Nisbet EC (2012) Boomerang effects in science communication: how motivated reasoning and identity cues amplify opinion polarization about climate mitigation policies. Commun Res 39.701–772. https://doi.org/10.1177/0093650211416646

Hart PS, Feldman L, Leiserowitz A, Maibach E (2015) Extending the impacts of hostile media perceptions: influences on discussion and opinion polarization in the context of climate change. Sci Commun 37(4):506–532. https://doi.org/10.1177/1075547015592067

Heath Y, Gifford R (2006) Free-market ideology and environmental degradation: The case of belief in global climate change. Environ Behav 38:48–71. https://doi.org/10.1177/0013916505277998

Hmielowski JD, Feldman L, Myers TA, Leiserowitz A, Maibach E (2014) An attack on science? Media use, trust in scientists, and perceptions of global warming. Public Underst Sci 23:866–883. https://doi.org/10.1177/0963662513480091

Hornsey MJ, Harris EA, Bain PG, Fielding KS (2016) Meta-analyses of the determinants and outcomes of belief in climate change. Nat Clim Chang 6:622–626. https://doi.org/10.1038/nclimate2943

Howe PD, Boudet H, Leiserowitz A, Maibach EW (2014) Mapping the shadow of experience of extreme weather events. Clim Chang:127, 381–389. https://doi.org/10.1007/s10584-014-1253-6

Iyengar S, Massey DS (2019) Scientific communication in a post-truth society. Proc Natl Acad Sci U S A 116:7656–7661. https://doi.org/10.1073/pnas.1805868115

Jacques PJ, Dunlap RE, Freeman M (2008) The organisation of denial: conservative think tanks and environmental scepticism. Environ Polit 17:349–385. https://doi.org/10.1080/09644010802055576

Joireman J, Barnes Truelove H, Duell B (2010) Effect of outdoor temperature, heat primes and anchoring on belief in global warming. J Environ Psychol 30:358–367. https://doi.org/10.1016/j.jenvp.2010.03.004

Kahan DM (2013) Making climate-science communication evidence-based – all the way down. SSRN J. https://doi.org/10.2139/ssrn.2216469

Kahan DM (2015) Climate-science communication and the measurement problem. Polit Psychol 36:1–43. https://doi.org/10.1111/pops.12244

Kahan DM, Jenkins-Smith H, Braman D (2011) Cultural cognition of scientific consensus. J Risk Res 14:147–174. https://doi.org/10.1080/13669877.2010.511246

Kahan DM, Peters E, Wittlin M, Slovic P, Ouellette LL, Braman D, Mandel G (2012) The polarizing impact of science literacy and numeracy on perceived climate change risks. Nat Clim Chang 2:732–735. https://doi.org/10.1038/nclimate1547

Kellstedt PM, Zahran S, Vedlitz A (2008) Personal efficacy, the information environment, and attitudes toward global warming and climate change in the United States. Risk Anal 28:113–126. https://doi.org/10.1111/j.1539-6924.2008.01010.x

Kerr JR (2020) Why do we argue about science? Exploring the psychological antecedents of rejection of science. Dissertation, Victoria University of Wellington. Available at: http://researcharchive.vuw.ac.nz/handle/10063/8640

Kim B, Kay DL, Schuldt JP (2021) Will I have to move because of climate change? Perceived likelihood of weather- or climate-related relocation among the US public. Clim Chang 165(9). https://doi.org/10.1007/s10584-021-03026-y

Kousky C, Kunreuther H, LaCour-Little M, Wachter S (2020) Flood risk and the U.S. housing market. J Hous Res 29:S3–S24. https://doi.org/10.1080/10527001.2020.1836915

Kraft PW, Lodge M, Taber CS (2015) Why people "don't trust the evidence": motivated reasoning and scientific beliefs. Ann Am Acad Polit Soc Sci 658:121–133. https://doi.org/10.1177/0002716214554758

Krosnick JA, Holbrook AL, Lowe L, Visser PS (2006) The origins and consequences of democratic citizens' policy agendas: a study of popular concern about global warming. Clim Chang 77:7–43. https://doi.org/10.1007/s10584-006-9068-8

Kulp SA, Strauss BH (2019) Author correction: new elevation data triple estimates of global vulnerability to sea-level rise and coastal flooding. Nat Commun 10:5752. https://doi.org/10.1038/s41467-019-13552-0

Kunda Z (1990) The case for motivated reasoning. Psychol Bull 108:480–498. https://doi.org/10.1037//0033-2909.108.3.480

Lazer DMJ, Baum MA, Benkler Y, Berinsky AJ, Greenhill KM, Menczer F, Metzger MJ, Nyhan B, Pennycook G, Rothschild D, Schudson M, Sloman SA, Sunstein CR, Thorson EA, Watts DJ, Zittrain JL (2018) The science of fake news. Science 359:1094–1109. https://doi.org/10.1126/science.aao2998

Leiserowitz A (2005) American risk perceptions: is climate change dangerous? Risk Anal 25:1433–1442. https://doi.org/10.1111/j.1540-6261.2005.00690.x

Leiserowitz A (2006) Climate change risk perception and policy preferences: The role of affect, imagery, and values. Clim Chang 77:45–72. https://doi.org/10.1007/s10584-006-9059-9

Leiserowitz A (2007) Communicating the risks of global warming: American risk perceptions, affective images, and interpretive communities. In: Moser SC, Dilling L (eds) Creating a climate for change: communicating climate change and facilitating social change. Cambridge University Press, pp 44–63. https://doi.org/10.1017/CBO9780511535871.005

Leiserowitz A, Maibach E, Roser-Renouf C, Feinberg G, Rosenthal S (2018) Climate change in the American mind. Yale University and George Mason University, New Haven

Levalle M (2018). Unable to bury climate report, Trump & deniers launch assault. Inside Climate News. Available at: https://insideclimatenews.org/news/30112018/fact-check-trump-climate-science-denial-national-assessment-sanders-global-warming/

Levy J, Bayes R, Bolsen T, Druckman JN (2021) Science and the politics of misinformation. In: Tumber H, Waisbord S (eds) Handbook of media, misinformation & populism. https://doi.org/10.4324/9781003004431

Lewandowsky S, Gignac GE, Vaughan S (2013) The pivotal role of perceived scientific consensus in acceptance of science. Nat Clim Chang 3:399–404. https://doi.org/10.1038/nclimate1720

Lodge M, Taber CS (2013) The rationalizing voter. Cambridge University Press, Cambridge. https://doi.org/10.1017/CBO9781139032490

Lorenzoni I, Nicholson-Cole S, Whitmarsh L (2007) Barriers perceived to engaging with climate change among the UK public and their policy implications. Glob Environ Chang 17:445–459. https://doi.org/10.1016/j.gloenvcha.2007.01.004

Ma Y, Dixon G, Hmielowski JD (2019) Psychological reactance from reading basic facts on climate change: the role of prior views and political identification. Environ Commun 13:71–86. https://doi.org/10.1080/17524032.2018.1548369

Malka A, Krosnick JA, Langer G (2009) The association of knowledge with concern about global warming: trusted information sources shape public thinking. Risk Anal 29:633–647. https://doi.org/10.1111/j.1539-6924.2009.01220.x

Marshall G (2015) Don't even think about it: why our brains are wired to ignore climate change. Bloomsbury Publishing, New York

McCright AM, Dunlap R (2000) Challenging global warming as a social problem: an analysis of the conservative movement's counter-claims. Soc Probl 47(4):499–522. https://doi.org/10.2307/3097132

McCright AM, Dunlap RE (2003) Defeating Kyoto: the conservative movement's impact on U.S. climate change policy. Soc Probl 50:348–373. https://doi.org/10.1525/sp.2003.50.3.348

McCright AM, Dunlap RE (2011) Cool dudes: the denial of climate change among conservative white males in the United States. Glob Environ Chang 21:1163–1172. https://doi.org/10.1016/j.gloenvcha.2011.06.003

Mildenberger M, Tingley D (2019) Beliefs about climate beliefs: the importance of second-order opinions for climate politics. Brit J Polit Sci 49:1279–1307. https://doi.org/10.1017/S0007123417000321

Mildenberger M, Lubell M, Hummel M (2019) Personalized risk messaging can reduce climate concerns. Glob Environ Chang 55:15–24. https://doi.org/10.1016/j.gloenvcha.2019.01.002

Milfont TL (2012) The interplay between knowledge, perceived efficacy, and concern about global warming and climate change: a one-year longitudinal study. Risk Anal 32:1003–1020. https://doi.org/10.1111/j.1539-6924.2012.01800.x

Morton TA, Rabinovich A, Marshall D, Bretschneider P (2011) The future that may (or may not) come: how framing changes responses to uncertainty in climate change communications. Glob Environ Chang 21:103–109. https://doi.org/10.1016/j.gloenvcha.2010.09.013

Moser S (2013) Successful adaptation to climate change: linking science and policy in a rapidly changing world, 1st edn. Routledge, Milton Park

Myers TA, Nisbet MC, Maibach EW, Leiserowitz AA (2012) A public health frame arouses hopeful emotions about climate change: a letter. Clim Chang 113:1105–1112. https://doi.org/10.1007/s10584-012-0513-6

Nisbet MC (2014) Engaging in science policy controversies. In: Bucci M, Trench B (eds) Routledge handbook of public communication of science and technology, pp 173–185. https://doi.org/10.4324/9780203483794.ch13

Nisbet MC, Myers T (2007) Polls trends: twenty years of public opinion about global warming. Public Opin Q 71:444–470. https://doi.org/10.1093/poq/nfm031

Nisbet EC, Cooper KE, Ellithorpe M (2015) Ignorance or bias? Evaluating the ideological and informational drivers of communication gaps about climate change. Public Underst Sci 24 (3):285–301. https://doi.org/10.1177/0963662514545909

Nyhan B, Reifler J (2010) When corrections fail: the persistence of political misperceptions. Polit Behav 32:303–330. https://doi.org/10.1007/s11109-010-9112-2

O'Connor RE, Bard RJ, Fisher A (1999) Risk perceptions, general environmental beliefs, and willingness to address climate change. Risk Anal 19:461–471. https://doi.org/10.1111/j.1539-6924.1999.tb00421.x

O'Neill SJ (2013) Image matters: Climate change imagery in US, UK and Australian newspapers. Geoforum 49:10–19. https://doi.org/10.1016/j.geoforum.2013.04.030

Oreskes N, Conway EM (2010) Merchants of doubt: how a handful of scientists obscured the truth on issues from tobacco smoke to global warming, 1st U.S. edn. Bloomsbury Press, New York

Osmundsen M, Bor A, Vahlstrup PB, Bechmann A, Petersen MB (2020) Partisan polarization is the primary psychological motivation behind "fake news" sharing on Twitter. PsyArXiv. https://doi.org/10.1017/S0003055421000290

Palm R, Bolsen T (2020) Climate change and sea-level rise in South Florida: the view of coastal residents. Springer Nature Switzerland AG, Cham. https://doi.org/10.1007/978-3-030-32602-9_2

Palm R, Bolsen T, Kingsland JT (2020) "Don't tell me what to do": resistance to climate change messages suggesting behavior changes. Weather Clim Soc 12:827–835. https://doi.org/10.1175/WCAS-D-19-0141.1

Pidgeon N, Fischhoff B (2011) The role of social and decision sciences in communicating uncertain climate risks. Nat Clim Chang 1:35–41. https://doi.org/10.1038/nclimate1080

Reser JP, Bradley GL, Ellul MC (2014) Encountering climate change: 'seeing' is more than 'believing'. WIREs Clim Change 5:521–537. https://doi.org/10.1002/wcc.286

Retchless DP (2018) Understanding local sea-level rise risk perceptions and the power of maps to change them: the effects of distance and doubt. Environ Behav 50:483–511. https://doi.org/10.1177/0013916517709043

Schuldt JP, Pearson AR (2016) The role of race and ethnicity in climate change polarization: evidence from a U.S. national survey experiment. Clim Chang 136:495–505. https://doi.org/10.1007/s10584-016-1631-3

Schuldt JP, Roh S (2014) Media frames and cognitive accessibility: what do "global warming" and "climate change" evoke in partisan minds? Environ Commun 8:529–548. https://doi.org/10.1080/17524032.2014.909510

Shao W, Keim BD, Garand JC, Hamilton LC (2014) Weather, climate, and the economy: explaining risk perceptions of global warming, 2001–10. Weather Clim Soc 6:119–134. https://doi.org/10.1175/WCAS-D-13-00029.1

Sheppard SRJ, Shaw A, Flanders D, Burch S, Wiek A, Carmichael J, Robinson J, Cohen S (2011) Future visioning of local climate change: a framework for community engagement and planning with scenarios and visualisation. Futures 43:400–412. https://doi.org/10.1016/j.futures.2011.01.009

Shukla JB, Verma M, Misra AK (2017) Effect of global warming on sea-level rise: a modeling study. Ecol Complex 32:99–110. https://doi.org/10.1016/j.ecocom.2017.10.007

Sinclair U (1994) I, Candidate for Governor: and how I got licked. University of California Press, Oakland

Slimak MW, Dietz T (2006) Personal values, beliefs, and ecological risk perception. Risk Anal 26:1689–1705. https://doi.org/10.1111/j.1539-6924.2006.00832.x

Smith N, Leiserowitz A (2012) The rise of global warming skepticism: exploring affective image associations in the United States over time. Risk Anal 32:1021–1032. https://doi.org/10.1111/j.1539-6924.2012.01801.x

Spence A, Pidgeon N (2010) Framing and communicating climate change: the effects of distance and outcome frame manipulations. Glob Environ Chang 20:656–667. https://doi.org/10.1016/j.gloenvcha.2010.07.002

Stern S (2005) Temporal dynamics of disclosure: the example of residential real estate conveyancing. Utah L Rev:57–95

Sundblad EL, Biel A, Gärling T (2007) Cognitive and affective risk judgements related to climate change. J Environ Psychol 27:97–106. https://doi.org/10.1016/j.jenvp.2007.01.003

Tesler M (2018) Elite domination of public doubts about climate change (not evolution). Polit Commun 35:306–326. https://doi.org/10.1080/10584609.2017.1380092

Tracey MD (2020) Don't fear flood data: here's how it helps clients. Realtor Magazine. Available at: https://magazine.realtor/sales-and-marketing/feature/article/2020/08/dont-fear-flood-data-heres-how-it-helps-clients

Tyson A, Kennedy B (2020) Two-thirds of Americans think government should do more on climate. Pew Research Center. Available at: https://www.pewresearch.org/science/2020/06/23/two-thirds-of-americans-think-government-should-do-more-on-climate/

Unsworth KL, Fielding KS (2014) It's political: how the salience of one's political identity changes climate change beliefs and policy support. Glob Environ Chang 27:131–137. https://doi.org/10.1016/j.gloenvcha.2014.05.002

van der Linden SL, Leiserowitz AA, Feinberg GD, Maibach EW (2014) How to communicate the scientific consensus on climate change: plain facts, pie charts or metaphors? Clim Chang 126:255–262. https://doi.org/10.1007/s10584-014-1190-4

van der Linden SL, Leiserowitz AA, Feinberg GD, Maibach EW (2015) The scientific consensus on climate change as a gateway belief: experimental evidence. PLoS ONE 10:e0118489. https://doi.org/10.1371/journal.pone.0118489

Weber EU, Stern PC (2011) Public understanding of climate change in the United States. Am Psychol 66:315–328. https://doi.org/10.1037/a0023253

Zaval L, Keenan EA, Johnson EJ, Weber EU (2014) How warm days increase belief in global warming. Nat Clim Chang 4:143–147. https://doi.org/10.1038/nclimate2093

Zhou J (2016) Boomerangs versus javelins: how polarization constrains communication on climate change. Environ Polit 25(5):788–811. https://doi.org/10.1080/09644016.2016.1166602

Zia A, Todd AM (2010) Evaluating the effects of ideology on public understanding of climate change science: how to improve communication across ideological divides? Public Understand Sci 19:743–761. https://doi.org/10.1177/0963662509357871

Chapter 5
The Role of the Real Estate Agent

Abstract Requirements for licensure as a real estate agent have become increasingly demanding over the past decades, involving the acquisition of knowledge and skills that make real estate agents invaluable to both home sellers and homebuyers. The National Association of Realtors® has set a stringent code of ethics for its members. In addition, Florida statutes require the disclosure of "all known facts that materially affect the value of residential property and are not readily observable." Empirical research has demonstrated the important influence that real estate agents have on buyers, and therefore an assessment of their evaluation of trends in the property market are crucial for an understanding of current practice and trends.

Keywords Realtor® · Disclosure · National Association of Realtors® (NAR) · Housing search behavior

Two million Americans hold real estate licenses. Almost anyone who has bought or sold a house has met with a real estate agent. But one might not appreciate the training of these real estate agents, or the history of how they came to hold the ethics and values they ascribe to.

Whether estate agents can be considered "professionals" is a topic that is still open for debate (Besbris 2017; Hornstein 2005). The classic definition of a profession (Otterlei 2018) includes "a long formal education, ethical guidelines, and great freedom and autonomy in their work." In real estate, unlike traditional professions such as law or medicine, there are no prescribed degree requirements for entry into the field. However, real estate associations have promoted stringent state licensing regulations that set minimum requirements to ensure that licensed agents follow state law and regulations. In this way, they raised the bar for entry into this field.

Real estate agents have a major influence on the search process for homebuyers, and even the eventual price at which the house will sell (Besbris 2020; Besbris and Faber 2017; Jauregui and Hite 2010; Jud and Frew 1986; Munneke et al. 2015). Although internet-based information about housing vacancies is now widely available, real estate agents have maintained their centrality in the purchase process. The National Association of Realtors® (2020) estimates that 88% of homebuyers

© The Author(s), under exclusive license to Springer Nature Switzerland AG 2022
R. Palm, T. Bolsen, *Housing Market Response to Sea-Level Rise in Florida*, Coastal Research Library 37, https://doi.org/10.1007/978-3-030-88435-2_5

purchased their home through a real estate agent or broker. Real estate agents not only deal with some of the details leading to the final contract between the seller and buyer, but also influence buyer perspectives on various neighborhoods and even the significance of homeownership itself in American society. In addition, agents are likely to link buyers with related real estate services including title companies, mortgage lenders and appraisers, as well as home-care specialists such as decorators, home cleaners, and gardeners or lawn services.

This chapter describes the development of the real estate agent profession and the emergence of their major trade association, the National Association of Realtors®. It examines the changing role of the real estate agent in the home purchase process, particularly with the evolution of the Multiple Listing Service and its development as a web-based information source. The state-mandated qualifications of a Florida real estate agent, as well as the general characteristics of the respondents to a 2020 survey of Florida real estate agents are summarized. Finally, the chapter outlines what the Florida respondents said about the significance of susceptibility to flooding in the location decision process.

5.1 The Real Estate Sales Profession

The Association of Real Estate License Law Officials estimates there are approximately two million active real estate licensees in the United States (Folger 2019). Of these, about 70% are "Realtors" or members of the National Association of Realtors®. This organization requires annual dues and members must agree to abide by their Code of Ethics and Standards of Practice.[1]

Founded in 1908, the history of the National Association of Realtors® (NAR), reflects the evolution of values surrounding the residential property market in American history. The initial goal of the predecessor organization – the National Association of Real Estate Boards – was the professionalization of the field with the goal of improving the reputation of board members as ethical businesspeople. Its task was to separate the reputation of board members from that of "fly-by-night swindlers" (Hornstein 2005). Women were excluded from the organization in its first decade, as the organization attempted to unify local boards into "a tenuous national brotherhood that served to unite men with disparate interests and experiences" (Hornstein 2005).

The association collaborated in developing the academic field of land economics, texts in the principles of real estate, and criteria for zoning and real estate appraisal that reflected the values of the times. For example, the appraisal values of the real estate boards were codified into the Federal Housing Administration guidelines

[1]The term "Realtor" is sometimes incorrectly used as a generic description for a real estate agent. Instead, the word should be capitalized, and refers only to members of the National Association of Realtors®.

enforcing what later became known as "redlining" or discouraging investment in portions of the city based on the ethnic and racial composition of its residents: Article 34 of Realtor Code of Ethics stated that the Realtor "should never be instrumental in introducing into a neighborhood a character of property or occupancy, members of any race or nationality, or any individuals whose presence will clearly be detrimental to property values in that neighborhood (National Association of Realtors® 2021a)." This portion of the code of ethics remained unchanged until 1950.

The National Association of Realtors® was also instrumental in the development of state licensure for real estate brokers, adding mandatory expertise and professionalism to the field. The first model licensing law, adopted in 1918, required that the applicant demonstrate sound character and required references: "an applicant needed to provide affidavits from two or more 'prominent' personages from his or her hometown who would vouch for the applicant's integrity and creditworthiness" (Hornstein 2005).

The association added a major asset to membership with the establishment of an information exchange, the Multiple Listing Service or MLS. This exchange was first developed in the 1980s as a systematic way to share profits from the selling of properties *(National Association of Realtors®* (2021b). The MLS compiled all the properties for sale listed for each board member, and when combined with a structure by which all participants in the sale could share commissions, the result was a system whereby "property listed with one is listed with all." After a short period of time when only the original listing agent sometimes had an exclusive right-to-sell agreement, board members were required to share all their listings with all other board members. It was obviously advantageous for real estate agents to have access to the MLS to obtain a wide range of properties that they could represent. Indeed, this membership advantage was so obvious that a series of lawsuits from the middle 1960s through the 1980s attempted to challenge the rule that only board members or Realtors® could have access to the MLS (National Association of Realtors® (2021c).

But until the 1950s, not all real estate agents could participate in the MLS. Local boards operated with their own bylaws that sometimes "explicitly stated that blacks, women, Jews, and other groups were not allowed to join" (National Association of Realtors® 2018). It was only in 1961 that the NAR overturned local board policies excluding African American real estate agents. African American real estate agents had formed their own local boards in major cities, and an association of local boards in 1947, the National Association of Real Estate Brokers (NAREB). Members were not called "Realtors®", but instead "Realtists®" (National Association of Real Estate Brokers 2021a). This organization still exists with 28 member chapters providing such services as "Realtist®-only listing opportunities in locations where NAREB chapters are active" (National Association of Real Estate Brokers 2021b). The total number of African Americans in the real estate profession remains low, comprising only 4.2% of all brokers and agents in the United States (Martin 2014).

The MLS evolved into an internet-based information system which continues to include an increasing amount of information and even virtual house tours. The website realtor.com was launched in 1995, and a year later became available to any user of the internet (Klein 2013).

5.2 Real Estate Agents in Florida

Each state sets requirements for the licensing of real estate brokers and associates. In Florida, statute defines a real estate broker as a person "who, for another, and for compensation or in anticipation of compensation, appraises, auctions, sells, exchanges, buys, rents or offers, attempts or agrees to appraise, auction, or negotiate any sale, exchange, purchase, or rental of business enterprises or any real property or any interest in or concerning the same" (State of Florida Government Statutes 475.01 2020). A broker associate is also licensed as a "broker" but "operates as a sales associate in the employ of another" (State of Florida Government Statutes 475.01 2020). The career path to broker involves both experience and mandated training. One enters the field of real estate in Florida as a "sales associate." A sales associate performs all the functions of a broker, but works "under the direction, control, or management of a broker or owner/developer" (State of Florida Government Statutes 475.01 2020). To be licensed as a sales associate, the individual must hold a high school diploma or equivalent, complete a 63-classroom hour pre-license education consisting of courses approved by the Florida Real Estate Commission (FREC) or hold a 4-year degree in real estate, and pass an examination. The requirements of "good character" have remained in the Florida Statutes as qualifications for practice: the applicant for a license must be "honest, truthful, trustworthy, and of good character; and have a good reputation for fair dealing" (State of Florida Government Statutes 475.17 2020). After receiving a sales associate license, the individual must take another 45 classroom hours of post-license education, a requirement waived for those with a 4-year degree or higher in real estate. Renewals are mandatory every two years, and each renewal at that level requires a further 14 classroom hours of continuing education.

To advance from sales associate to broker or broker-associate, the individual must have completed 72 approved classroom hours, certified by passage of an examination administered by an accredited college or real estate school. The individual must have held an active sales associate's license for "at least 24 months during the preceding 5 years in the office of one or more real estate brokers" (State of Florida Government Statutes 475.17 2020). After the license as broker is awarded, there are further requirements to renew the license: for the first renewal, an additional 60 approved classroom hours, and after the first renewal, 14 approved classroom hours of continuing education.

In Florida, there are three categories of broker: principal broker, managing broker and associate broker. The principal broker oversees all the licensed real estate agents in the firm, ensuring their compliance with state and national law (Bortz 2020). There may be a managing broker who oversees the office and an associate broker who also holds a broker's license but is supervised by a managing broker and does not have supervisory responsibilities.

The Florida Realtors®, the state affiliate of the National Association of Realtors®, began in 1911 with 61 members. In 2020, with almost 200,000 active members, Florida is second only to California in numbers of Realtors®, and first in the nation in Realtors® per capita (National Association of Realtors® 2019).

5.3 How Real Estate Agents Influence the Search for a New Home

Having described the development of the residential real estate sales industry in the United States and in Florida, we turn to the question of how real estate agents affect the home-search behaviors of prospective buyers. We might first ask what prospective buyers are seeking when they search for a new home. When we posed this question, several of the survey respondents wisely reminded us that there is no generic buyer, and therefore no general answer. Instead, this decision is affected by different sets of experiences, constraints, available resources, and preferences with respect to the structure itself, its setting, and the general area in which it is located. The decision is made within important resource limitations, and at a particular time and place, requiring trade-offs between what might be desired, what is available, and what the buyers wish to pay.

Generally, economic constraints set the most important boundaries for the home search. The household must determine the resources it can and wishes to invest in housing, including an assessment of the cost and availability of a home mortgage loan, and other costs associated with homeownership such as property taxes, special assessments, and homeowners' association fees. Additional costs include those for the upkeep of the structure itself and the property around it, as well as costs associated with locational trade-offs such as costs of commuting to job(s), schools, and other activities in terms of both time and actual monetary expense.

With respect to the structure, preferences vary, but regular surveys of recent home-purchasers by the National Association of Home Builders (NAHB 2019) reveal something current of general trends: recent buyers have reported preferring houses with garage storage, walk-in pantries, hardwood flooring, a patio, and exterior lighting, as well as smaller overall size for the house and garage, kitchens with stainless steel appliances and natural stone countertops, and wireless controls. More generally, a Zillow survey (McLaughlin and Young 2017) found that the most important factors driving home selection were that the property was within the buyer's budget, that it is in a safe neighborhood, it has a floor plan that meets the household's needs and has the preferred number of bedrooms. Less important were community amenities, proximity to transit, storage, a view, or off-street parking.

The next decision the prospective home buyer needs to make is that of location or neighborhood. Despite the sporadic adoption of laws and ordinances to permit "upzoning" or the construction of higher density residences within neighborhoods previously zoned for single family dwellings (Manville et al. 2020; Rodriguez-Pose and Storper 2020; Wegmann 2020), survey research reports a persistent preference within the United States for single-family, detached dwellings at relatively low densities (Audirac and Smith 1992; Brown et al. 1997; Jansen 2020; Gehrke et al. 2019; Anderson 2019).

The home search process itself has been modelled to account for the sequence of decisions the household must make and the very real constraints to learning about the actual availability of housing. Bruch and Swait (2019) proposed a "multistage neighborhood choice process" that first involves the decision of whether to stay where they are or move, a decision that is often affected by changes in household size or resources. If the decision is to try to move, the household must then decide on a limited number of neighborhoods to consider, usually circumscribed by a combination of geographic constraints and housing affordability (Huff 1986; Piazzesi et al. 2015; Smith and Clark 1982). The search process is also constrained by the limited amount of time the household might have to find a home, and the fact that it may not be possible to learn about all available vacancies (Bruch and Swait 2019; Dunning and Grayson 2014).

Before the advent of competing websites advertising properties for sale, the real estate agent was a kind of gate keeper in providing information to prospective homebuyers. Indeed, the real estate agent had been described as a kind of "geography teacher" in selecting, editing, and conveying information about the city (Palm 2014). In addition to holding the information contained in the Multiple Listing Service files, the real estate agent knew relatively more about the housing market than the clients and could therefore exert an important influence on where and how long the prospective buyers searched (Barresi 1968). Real estate agents could make the house search more efficient by reducing unnecessary visits to inappropriate vacancies, but also could engage in the well-documented practice of racial steering (Baryla and Ztanpano 1995; Elder et al. 1999; Galster and Godfrey 2005; Ondrich et al. 2003; Zumpano et al. 2003).

With the conversion of information about housing vacancies from a card file to a paperbound book and then to a website, prospective buyers gained the ability to participate in the selection more actively. Although there was uneven uptake of internet information on housing vacancies in the early years (Littlefield et al. 2000; Palm and Danis 2003), there is now heavy reliance on this source of information (Cherif 2013; Dunning and Grayson 2014; McLaughlin and Young 2017). Initial fears in the industry that the availability of such data would result in a decline in the use of real estate agents (Baen and Guttery 1997) did not come to fruition, and instead, buyers who consult online sources also continue to depend on real estate agents (McLaughlin and Young 2017). The continued reliance of prospective buyers on real estate agents is partly explained by the significance and complexity of the home purchase process, in comparison with the process of for example, booking a flight, hotel, or vacation, which can be completed entirely online and represents only a fraction of the cost. The stakes are high for households seeking a new home, and they continue to depend on a personal contact with real estate agents for guidance in making this important decision (Adams 2020; Grabar 2018; Tracey 2019).

5.4 The Duty to Disclose by the Seller and the Real Estate Agent

Among the responsibilities of the real estate agent is the duty to disclose certain features of the property that might not appear in online advertisements or even be visible to prospective buyers in an on-site visit. This requirement varies from one state to another and reflects an evolution in common law as well (Dallon 2002; Lefcoe 2004). In the 1950s, sellers and their agents "had no obligation to mention property defects to buyers" as long as they did not outright lie about them (Lefcoe 2004). This left the purchase process as a kind of *caveat emptor* or "buyer beware." However, by the mid-1960s, statutes in most states required seller disclosure of known material defects.

In 1985, following a court decision finding a listing broker liable for damage to a California home caused by landslides, the California Association of Realtors® developed guidelines suggesting disclosure of property defects to homebuyers in writing, a process that was put into state law in 1986. Instead of placing the legal responsibility for disclosure with the real estate agent, professional real estate associations pressed for responsibility for the disclosure to rest with the seller, and over the next few years about two-thirds of the states had mandated disclosure by the seller (Lefcoe 2004). These disclosure laws may include risk for fire, land movement or earthquake damage, and flooding, although again the specific laws and the disclosure forms required vary by state.

As in other states, Florida mandates that the real estate licensee disclose to both the buyer and seller "all known facts that materially affect the value of residential property and are not readily observable" (State of Florida Government Statutes 475.278 2020). Significantly, however, Florida is one of 21 states with "no statutory or regulatory requirements for a seller to disclose a property's flood risks or past flood damages to a potential buyer" (Scata 2018). The Florida Association of Realtors® has developed a standard form for **voluntary** disclosure including questions about whether there is past or present water intrusion, drainage, or flooding problems, if it is seaward of the coastal construction control line, if it has an elevation certificate, and whether the lender requires flood insurance.

In 2020, a bill was introduced into the Florida Senate (SB1842), mandating that sellers disclose knowledge of existing flood insurance coverage, location of the property in a floodplain or reservoir, and any previous flood damage. However, the bill never became law.[2] In Florida, county-level disclosure mandates exist, but there is no uniform state-wide rule. Instead, Barefoot et al. (2020) describe Florida's current common law as one that "effectively expects buyers to perform inspections beyond what is customary, to research and analyze flood maps and records, and to

[2] Florida statutes do, however, contain a provision requiring sellers to provide a written disclosure at the time of contract if the property is seaward of the coastal construction control line, indicating that it may be subject to beach nourishment and the protection of marine turtles (Florida Statutes 161.57).

knock on the doors of their prospective neighbors in order to evaluate their flood risk."

5.5 Are Florida Homebuyers Concerned with Flood Risk?

Given the combination of increased high tide flooding in coastal Florida (Hino et al. 2019; Sweet et al. 2017; Wdowinski et al. 2016) and the absence of a mandatory flood risk disclosure statute, how do real estate agents in South Florida rank susceptibility to flooding among the considerations they believe that buyers consider when searching for a home? We presented the survey respondents with ten neighborhood qualities and asked them to indicate the importance of each in the search process. The wording of the question was:

> As a Florida real estate professional, we know that you are familiar with local housing market trends. We are interested in your opinion as to how buyers decide on neighborhoods in which to search for houses. In your experience, how important are the following factors for buyers when deciding on a neighborhood in which to purchase a house?

Ten factors were listed, and the respondents were asked to choose a level of importance from 1 (not at all important) to 7 (most important). The ten neighborhood characteristics were included in the questionnaire were: age of the neighborhood, noise levels in the neighborhood, parking/traffic in the neighborhood, presence of children in the neighborhood, proximity to a beach, park or playground, proximity to work or commute time, quality of the local public schools, safety from crime in the neighborhood, safety from flooding in the neighborhood and proximity to restaurants, shops, and entertainment.

Overall and in every submarket, real estate agents evaluated safety from crime in the neighborhood as the most important factor from this list (Table 5.1). Second was "quality of the local public schools", which is also interpreted, even for households without school children, as a general indicator of the socioeconomic status of the neighborhood. Least important were the general age of the neighborhood or the presence/absence of children.

Safety from flooding ranked as #3 out of 10 overall but submarket rankings ranged from #2 in Fort Myers-Cape Coral to #8 in West Palm. The range of housing available within the submarket could have affected the variability these rankings of the significance of freedom from flooding, in that, as Fu and Nijman have noted (2021), housing at greater range of elevations is available in places such as Tampa-St. Petersburg and Fort Myers than in the Atlantic coastal locations including West Palm and Miami.

We also asked the real estate agents to enumerate other factors that they thought were important neighborhood characteristics. Among the common responses were neighborhood affordability, amenities such as swimming pools, golf courses, and boating access, estimated future value of properties in the neighborhood, homeowner association fees, walkability, and the presence of gates. The most

Table 5.1 Real estate agent rankings of the relative importance of neighborhood characteristics for homebuyers

	Crime	Schools	Flooding	Shops	Beach	Commute	Noise	Parking	Kids	Age	**Relative rank of flooding**
All	5.2	4.5	4.3	4.1	4	4	4	3.9	3	2.9	**3**
Tampa	5.1	4.5	4.5	4.3	4.1	4.2	3.9	3.8	2.8	2.6	**2.5**
Ft Myers	5.2	3.8	4.4	4.2	4.1	3.4	4.3	3.8	2.8	2.8	**2**
Ft Laud.	5.3	4.9	4.1	3.9	3.7	4.4	3.9	4	3.1	3.1	**4**
Miami	5.3	4.9	4	4.3	3.9	4.4	4.1	4.4	3.2	3.2	**7**
West Palm	5.5	4.7	3.8	4	4	3.9	4.2	3.9	3.4	3.4	**8**
North Port	5	4.3	4.5	4.5	4.6	3.7	4.1	3.8	2.8	2.8	**3.5**

frequent response, however, was the admonition that buyers vary markedly with respect to what they value when seeking a new home.

Of interest is the comparison of this preference list to that which we obtained from homeowners surveyed in 2018 (Palm and Bolsen 2020). We posed the question "how important were the following factors to you as you decided where you wanted to live". We provided five choices (schools, commute, shops, flooding, and crime) as well as "other." Like the real estate agents, homeowners reported that safety from crime in the neighborhood was by far the most important factor in their location decision. Unlike the real estate agents' response, however, quality of the local public schools was ranked last, an evaluation which was no doubt related to the fact that the average age of the owners of single-family dwellings in our survey was 51.2 years. Other factors mentioned most frequently by the residents as important in their home search were price/affordability and general quality of the neighborhood.

5.6 Summary

The real estate profession evolved in the twentieth century with the active participation of the National Association of Realtors® defining ethical standards and business practices. This organization had been small and exclusive, limited to white men. However, it now reaches out to all residential real estate licensees who wish to apply for membership. A majority of the members are now women, although the organization still is not as diverse ethnically as the general population.

The code of ethics of the NAR guided the nature of property value appraising in the past, and currently sets high standards for the sharing of information with prospective buyers. It has lobbied for the adoption of state laws mandating extensive seller disclosure of property characteristics. The State of Florida also has developed demanding standards for real estate licensees, although the state does not currently mandate the disclosure of information about the past-flooding or flood susceptibility of individual properties.

The real estate agent maintains a central position in the home purchase process, a position that has been supplemented but not replaced by the widespread availability of information about housing vacancies on the internet. Real estate agents have an important influence on home buyer understanding of neighborhoods, and their opinions on the housing market are therefore very important. The real estate agents we surveyed themselves saw safety from flooding as one of the more important criteria that prospective buyers consider, although far less significant than safety from crime or quality of the local public schools. With this background, we are now ready to interpret how real estate agents in Florida view sea-level rise as a factor in the current housing market.

References

Adams BE (2020) Why Silicon Valley has failed to replace your real estate agent (so far). In: Hood Homes Blog. Available at: https://www.hoodhomesblog.com/real estate-industry/silicon-valley-has-failed-to-replace-real estate-agents/

Anderson D (2019) Millennials still want single-family homes even if it means a long commute. In: Redfin News. Available at: https://www.redfin.com/news/millennial-homebuyers-prefer-single-family-homes/

Audirac I, Smith MT (1992) Urban form and residential choice: preference for urban density in Florida. J Archit Plan Res 9(1):19–32

Baen J, Guttery R (1997) The coming downsizing of real estate: implications of technology. J Real Estate Portf Manag 3(1):1–18. https://doi.org/10.1080/10835547.1997.12089535

Barefoot N, Fleming A, Tagtachian D, Falla G, Blackeman B, Cavellier (2020) There will be floods: armoring the people of Florida to make informed decisions on flood risk. Fla Bar J 94(5):28–32

Barresi CM (1968) The role of the real estate agent in residential location. Sociol Focus 1(4):59–71. https://doi.org/10.1080/00380237.1968.10570511

Baryla E, Ztanpano L (1995) Buyer search duration in the residential real estate market: the role of the real estate agent. J Real Estate Res 10:1–13. https://doi.org/10.1080/10835547.1995.12090769

Besbris M (2017) The interactional market: real estate agents, homebuyers, and the production of value. Dissertation, New York University

Besbris M (2020) Upsold: real estate agents, prices, and neighborhood inequality. University of Chicago Press, Chicago. https://doi.org/10.7208/chicago/9780226721408.001.0001

Besbris M, Faber JW (2017) Investigating the relationship between real estate agents, segregation, and house prices: steering and upselling in New York State. Sociol Forum 32(4):850–873. https://doi.org/10.1111/socf.12378

Bortz D (2020) Real estate agent, broker, realtor: what's the difference? In: Realtor.com. Available at: https://www.realtor.com/advice/buy/whats-difference-real estate-salesperson-broker/

Brown DL, Fuguitt GV, Heaton TB, Waseem S (1997) Continuities in size of place preferences in the United States, 1972–1992. Rural Sociol 62(4):408–428. https://doi.org/10.1111/j.1549-0831.1997.tb00657.x

Bruch E, Swait J (2019) Choice set formation in residential mobility and its implications for segregation dynamics. Demography 56(5):1665–1692. https://doi.org/10.1007/s13524-019-00810-5

Cherif E (2013) Analysis of the internet impact on the real estate industry. Int J Serv Sci Manag Eng Technol 4(3):51–67. https://doi.org/10.4018/ijssmet.2013070104

Dallon CW (2002) Theories of real estate broker liability and the effect of the as is clause. Fla L Rev 54:395

Dunning RJ, Grayson A (2014) Homebuyers and the representation of spatial markets by information providers. Int J Hous Mark Anal 7(3):292–306. https://doi.org/10.1108/IJHMA-07-2013-0045

Elder HW, Zumpano LV, Baryla EA (1999) Buyer search intensity and the role of the residential real estate broker. J Real Estate Financ Econ 18(3):351–368. https://doi.org/10.1023/A:1007737102125

Florida Senate (2020) An act relating to residential property disclosures: SB 1842. https://www.flsenate.gov/Session/Bill/2020/1842/BillText/Filed/PDF

Folger J (2019) How to become a real estate agent. In: Millionacres: a Motley fool service. Available at: https://www.fool.com/millionacres/real estate-basics/how-become-real estate-agent/

Fu X, Nijman J (2021) Sea-level rise, homeownership, and residential real estate markets in South Florida. Prof Geogr 73(1):63–71. https://doi.org/10.1080/00330124.2020.1818586

Galster G, Godfrey E (2005) By words and deeds: racial steering by real estate agents in the US in 2000. J Am Plan Assoc 71(3):251–268. https://doi.org/10.1080/01944360508976697

Gehrke SR, Currans KM, Clifton KJ (2019) Assessing the importance of housing, accessibility, and transportation characteristics on stated neighbourhood preference. Int J Urban Sci 23(1):49–66. https://doi.org/10.1080/12265934.2018.1436983

Grabar H (2018) Why aren't we all buying houses on the internet? In: Slate. Available at: https://slate.com/technology/2018/06/redfin-zillow-and-opendoor-cant-disrupt-real estate-agents-and-may-not-want-to.html

Hino M, Belanger ST, Field CB, Davies AR, Mach KJ (2019) High-tide flooding disrupts local economic activity. Sci Adv 5(2):eaau2736. https://doi.org/10.1126/sciadv.aau2736

Hornstein JM (2005) A nation of Realtors®: a cultural history of the twentieth-century American middle class. Duke University Press, Durham. https://doi.org/10.2307/j.ctv11sn0rc

Huff JO (1986) Geographic regularities in residential search behavior. Ann Assoc Am Geogr 76 (2):208–227. https://doi.org/10.1111/j.1467-8306.1986.tb00113.x

Jansen SJT (2020) Urban, suburban or rural? Understanding preferences for the residential environment. J Urban Int Res Placemaking Urban Sustain 13(2):213–235. https://doi.org/10.1080/17549175.2020.1726797

Jauregui A, Hite D (2010) The impact of real estate agents on house prices near environmental disamenities. Hous Policy Debate 20(2):295–316. https://doi.org/10.1080/10511481003738419

Jud GD, Frew J (1986) Real estate brokers, housing prices, and the demand for housing. Urban Stud 23(1):21–31. https://doi.org/10.1080/00420988620080031

Klein S (2013) Early history of realtor.com: a chronology of events. In: Data Advocate. Available at: https://www.thedataadvocate.com/early-history-of-realtor-com-a-chronology-of-events/

Lefcoe G (2004) Property condition disclosure forms: how the real estate industry eased the transition from caveat emptor to seller tell all. Real Prop Prob Tr J 39:193–250

Littlefield JE, Bao Y, Cook DL (2000) Internet real estate information: are home purchasers paying attention to it? J Consum Mark 17(7):575–590. https://doi.org/10.1108/07363760010357787

Manville M, Monkkonen P, Lens M (2020) It's time to end single-family zoning. J Am Plan Assoc 86(1):106–112. https://doi.org/10.1080/01944363.2019.1651216

Martin ST (2014) In ranks of selling real estate, a scarcity of black realtors and brokers. In: Tampa Bay Times, 1 Sep 2014. Available at: https://www.tampabay.com/news/business/realestate/in-ranks-of-selling-real estate-a-scarcity-of-black-realtors-and-brokers/2195692/

McLaughlin R, Young C (2017) A shared future: data democratization and spatial heterogeneity in the housing market. Paper presented at the Harvard Joint Center for Housing Studies symposium April 2017. Available at: https://www.jchs.harvard.edu/sites/default/files/media/imp/a_shared_future_data_democratization_spatial_heterogeneity.pdf

Munneke HJ, Ooi JT, Sirmans CF, Turnbull GK (2015) Real estate agents, house prices, and liquidity. J Real Estate Financ Econ 50(1):1–33. https://doi.org/10.1007/s11146-013-9452-x

National Association of Home Builders (2019) NAHB outlines top features homebuyers want for 2019. Available at: https://www.nahb.org/News-and-Economics/Industry-News/Press-Releases/2019/02/NAHB-Outlines-Top-Features-Home-Buyers-Want-for-2019

National Association of Real Estate Brokers (2021a) Membership services. Available at: https://www.nareb.com/membership-services/

National Association of Real Estate Brokers (2021b) NAREB history. Available at: https://www.nareb.com/our-history

National Association of Realtors® (2018) From one voice to many. Available at: https://www.nar.realtor/sites/default/files/documents/March-2018-Fair-Housing-3-9-2018.pdf. Accessed 21 Mar 2021

National Association of Realtors® (2019) Historic membership county by state. Available at: https://www.nar.realtor/membership/historic-report. Accessed 23 Feb 2021

National Association of Realtors® (2020) Quick real estate statistics. Available at: https://www.nar.realtor/research-and-statistics/quick-real-estate-statistics

National Association of Realtors® (2021a) Multiple Listing Service (MLS): what is it. Available at: https://www.nar.realtor/nar-doj-settlement/multiple-listing-service-mls-what-is-it Accessed 15 Jan 2021

National Association of Realtors® (2021b) 1924 Code of Ethics. Available at: https://www.nar.realtor/about-nar/history/1924-code-of-ethics Accessed 28 Oct 2020

National Association of Realtors® (2021c) Non-member access to REALTOR® Association Multiple Listing Services. Available at: https://www.nar.realtor/legal/non-member-access-to-realtor-association-multiple-listing-services Accessed 15 Jan 2021

Onrich J, Ross S, Yinger J (2003) Now you see it, now you don't: why do real estate agents withhold available houses from Black customers? Rev Econ Stat 85(4):854–873. https://doi.org/10.1162/003465303772815772

Otterlei JB (2018) What is a profession, and what are the prerequisites for being a profession? In: Örtenblad A (ed) Professionalizing leadership. Palgrave Macmillan, Cham. https://doi.org/10.1007/978-3-319-71785-2_2

Palm R (2014) A personal history. In: Berry BJL, Wheeler J (eds) Urban geography in America, 1950–2000. Paradigms and personalities. Routledge, New York, pp 139–145. https://doi.org/10.4324/9781315880952

Palm R, Bolsen T (2020) Climate change and sea-level rise in South Florida. Springer, Cham. https://doi.org/10.1007/978-3-030-32602-9

Palm R, Danis M (2003) The internet and home purchase. Tijdschr Econ Soc Geogr 93(5):537–547. https://doi.org/10.1111/1467-9663.00224

Piazzesi M, Schneider M, Stroebel J (2015) Segmented housing search (No. w20823). National Bureau of Economic Research. doi:https://doi.org/10.3386/w20823

Rodríguez-Pose A, Storper M (2020) Housing, urban growth and inequalities: the limits to deregulation and upzoning in reducing economic and spatial inequality. Urban Stud 57 (2):223–248. https://doi.org/10.1177/0042098019859458

Scata J (2018) Homebuyers face stacked deck to learn of past floods. In: NDRC Expert Blog,16 Aug 2018. Available at: https://www.nrdc.org/experts/joel-scata/home-buyers-face-stacked-decks-learn-past-floods

Smith TR, Clark WA (1982) Housing market search behavior and expected utility theory: 1. Measuring preferences for housing. Environ Plan A 14(5):681–698. https://doi.org/10.1068/a140681

Sweet WWV, Kopp R, Weaver CP, Obeysekera JTB, Horton RM, Thieler ER, Zervas CE (2017) Global and regional sea-level rise scenarios for the United States. NOAA Technical Report NOS CO-OPS-83. https://doi.org/10.7289/v5/tr-nos-coops-083

Tracey MD (2019) You're not being replaced by AI. In: Realtor® Magazine Live, 16 May 2019. Available at: https://magazine.realtor/daily-news/2019/05/16/you-re-not-being-replaced-by-ai

Wdowinski S, Bray R, Kirtman BP, Wu Z (2016) Increasing flooding hazard in coastal communities due to rising sea-level: case study of Miami Beach, Florida. Ocean Coast Manag 126:1–8. https://doi.org/10.1016/j.ocecoaman.2016.03.002

Wegmann J (2020) Death to single-family zoning… and new life to the missing middle. J Am Plan Assoc 86(1):113–119. https://doi.org/10.1080/01944363.2019.1651217

Zumpano LV, Johnson KH, Anderson RI (2003) Internet use and real estate brokerage market intermediation. J Hous Econ 12(2):134–150. https://doi.org/10.1016/S1051-1377(03)00018-4

Chapter 6
Methodology: Survey of South Florida Realtors

Abstract The authors collected survey responses from 680 licensed real estate agents in south Florida in the fall of 2020. Compared with national statistics, the Florida respondents were more likely to be male, have completed higher levels of education, were older, and had more years of experience in real estate. Politically, most identified themselves as Independents, and higher percentages of Gulf Coast real estate agents said they were Republicans. Almost all of them believe that the sea-level is rising, that climate change is occurring, and that sea-level rise is a product of climate change.

Keywords Realtors® · Submarkets · Survey research

From September through November 2020, we administered an internet-based survey to brokers and real estate agents with office addresses in 13 counties in south Florida.[1] We received responses from 330 brokers and 350 sales associates, all of whom were affiliated with the Florida Association of Realtors®. The names and email addresses of these agents were obtained from the "Email List Company," one of many companies that compile listings of email addresses of various occupational groups to sell to marketers. The original list, dated 2020, included the names, office addresses, email addresses, phone numbers, real estate board membership and license information for 310,253 Florida real estate agents. Since the National Association of Realtors® monthly membership report for November 2020 notes that the Florida membership is only 195,747, it is likely that this list included numerous individuals who were retired or who no longer were engaged in real estate sales. In fact, several of the real estate agents we contacted responded that they had retired several years ago or were in commercial rather than residential real estate. The National Association of Realtors® estimates that about one-third of its membership does commercial rather than residential real estate exclusively. These individuals sell shopping centers or large real estate projects rather than single-family

[1] The questionnaire is Appendix 1 and the results of a "survey experiment" included in the design is described as Appendix 2.

© The Author(s), under exclusive license to Springer Nature Switzerland AG 2022

R. Palm, T. Bolsen, *Housing Market Response to Sea-Level Rise in Florida*, Coastal Research Library 37, https://doi.org/10.1007/978-3-030-88435-2_6

dwellings, and do not advise prospective homebuyers or sellers. Another factor that affects the delimitation of a survey population is the fact that the real estate profession has high turnover rates, particularly in the early ranks. Among sales associates, many people enter and leave the field annually and also change offices: nationally, almost one-fifth of the field have been licensed for a year or less, and the average tenure of a Realtor® with their current firm was just 4 years. All of this is to say that as researchers, we could not pinpoint the full population of real estate agents currently active and specializing in residential sales from which to draw a sample, but instead accepted the responses we received as a reasonable approximation of Florida real estate license holders who are classified as Realtors®.

We reduced the original list by selecting names from the following 13 south Florida counties: Monroe, Miami-Dade, Broward, Palm Beach, Lee, Charlotte, Sarasota, Manatee, Hillsborough, Pinellas, Indian River, St. Lucie and Martin. We then eliminated duplicates, all listings that did not show a business office, and any individual who did not have a real estate license listed with their identification. The list of 310,253 was reduced in this way to 15,388 brokers or "broker sales" and 80,059 "sales associates". The cover letter inviting the real estate brokers to respond promised a $10 Amazon gift card for completing the survey, and the letter to the sales associates promised a $1.00 gift card.

We administered surveys of real estate brokers and sales associates sequentially, since individuals in these categories have different levels of experience, different licensing requirements and different levels of responsibility. Over the course of October 19 – November 4, we emailed surveys to the list of all 15,388 brokers, and from November 16–23, to the list of 80,059 sales associates. Individuals who had not contacted us to "opt out" and who had not submitted the survey were recontacted twice after 3 days and after 7 days to remind them to complete the survey.

The final respondent pool consisted of 330 brokers and 350 sales associates. With a total of 680 high quality responses, including open-ended answers to questions we asked, we believe that we have a strong basis on which to draw conclusions about the real estate agent view of the South Florida housing market.

We carefully worded all questions and instructions to ensure that the respondents could understand the items on the questionnaire (Appendix 1). In addition, we tried to avoid politicized language by not linking the study in any way to "climate change" or "global warming". Although we did ask about the individually held beliefs concerning the existence of climate change, sea-level rise and the relationship between the two near the end of the questionnaire, we avoided linking these potentially emotion-charged concepts with our examination of the real estate market itself. Instead, our questions used the terms "risk of flooding" and "elevation above sea-level." These are also precisely the terms that have been used in the econometric studies positing declines in coastal house prices or price trends.

After welcoming the respondents to the survey, they were directed to the consent form which described the purpose of the study: "to understand the housing market from the perspective of those professionals who work most closely with it: real estate agents." We noted that we were studying the "experience of real estate agents in south Florida, given the risk of sea-level rise."

The first section of the survey was a set of questions asking about the length of their experience in real estate, how long they had lived in Florida, the geographic region, their age, sex, race or ethnicity, education, political party identification and the geographic location of the primary real estate market they cover. The next set of questions asked them to evaluate the importance of various factors that buyers consider when deciding on a neighborhood in which to purchase a house. We presented ten possible factors in a randomized order, including the presence of children, noise levels, parking or traffic, proximity to work, proximity to a beach, park or playground, proximity to restaurants, shops and entertainment, safety from crime, safety from flooding, age of the neighborhood, and quality of the local public schools. We also provided a space for them to enter "other" factors.

The next section of the survey dealt with issues of low-lying or flood-prone areas. We asked whether they believed buyers are avoiding property in low-elevation coastal areas because of possible flooding, and to describe in their own words why people are or are not avoiding such properties. We reasoned that if some buyers are avoiding low-elevation coastal areas, this would reduce overall demand and therefore reduce prices. We asked whether they have observed house prices falling or not rising as rapidly for properties at risk of flooding compared with housing not susceptible to coastal flooding. We also asked whether the agents had noticed differences between buyers expecting to live in the property year-round as opposed to those either buying a vacation home or buying the property as an investment.

Another potential influence on house prices is the way in which appraisers or mortgage lenders respond to low-lying areas. We asked the real estate agents if they had observed mortgage loan applications being declined or requiring additional charges (points or mortgage insurance, for example) for houses in such areas. We also asked whether they had observed appraisers considering elevation or the likelihood of flooding in their assessment of home value. Finally, we noted that elevation might matter more at either the high end or the low end of the market, and we asked them whether they had observed differences in the significance of elevation for high priced as opposed to lower priced housing.

We next inquired about the response from either buyers or sellers to the new information about both Flood Factor and the FEMA flood zone, and whether the response had been positive or negative. We inquired about their own opinion concerning the decision of the National Association of Realtors® to add this information to all property listings on the realtor.com website, whether they would recommend that prospective buyers view this new information about property listings, and any other comments.

Finally, we asked them to speculate about the 5–10-year future of residential property in low elevation coastal areas: that is, would they see less demand and lower prices in areas prone to flooding. We concluded the survey with a battery of three questions aimed at their own attitudes about climate change and sea-level rise: whether they believe that the sea-level is rising, that climate change is happening and if sea-level rise is related to climate change.

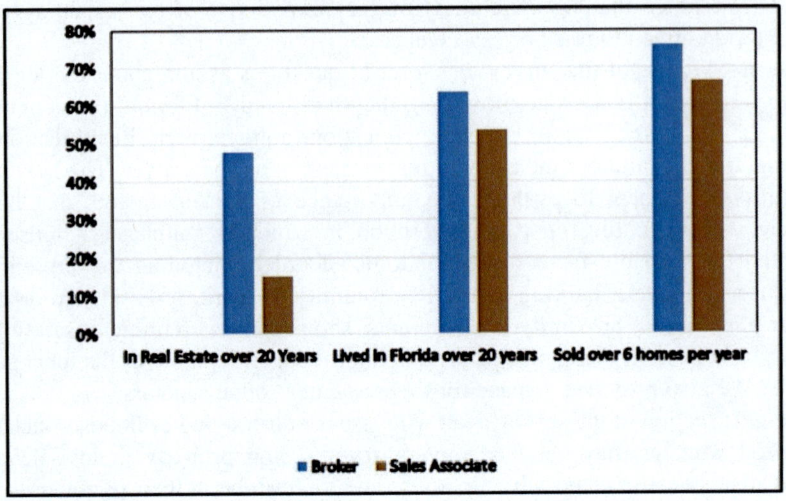

Fig. 6.1 Experience, residency and sales volume

6.1 Respondent Characteristics

With respect to experience, age and demographic characteristics, the Florida respondents were similar in several ways to national average as reported the National Association of Realtors®. This organization reports that 65% of Realtors® are women, that their median age is 54. Nationally, 22% have broker licenses and another 15% have broker associate licenses (Linsell 2020). Brokers are more likely than sales associates to be male, and the average Realtor® has nine years of experience. Because our respondents included more brokers than the national proportion, they were more likely to be male and to have had more experience in the profession. Like many of their clients, the Florida real estate agents we surveyed also tended to be older than the national average: 62% of the brokers and 56% of the sales associates were over age 55.

The Florida respondents were also very experienced in real estate and familiar with the state of Florida. Almost half of the brokers (48%) had been in the real estate profession for more than 20 years, and almost two-thirds had been Florida residents for more than 20 years. Sales associates had a median of 6–10 years in the real estate profession, and a majority had lived in Florida for more than 20 years (Fig. 6.1). More than two-thirds of the respondents reported selling more than 6 houses per year. Clearly, these respondents are familiar with the profession and with the areas in which they are selling real estate.

Nationally, relatively few persons of color sell real estate, certainly less than their proportions in the population, and these statistics hold for our sample of real estate agents in Florida as well. Fewer than 3% of our survey respondents reported themselves as being Black, even though Blacks make up around 17% of the Florida population (United States Census Bureau 2019). Asian Americans account for

3.0% of the Florida population, but only 1% of the respondents were Asian American. Although there was a stronger representation of "Hispanics" in the survey population, these numbers still do not reflect the total Florida population: more than 25% of Floridians are Hispanic,[2] but only 13% of the respondents in our study were Hispanic.

6.2 Submarket Differences

As we noted in Fig. 1.5, we divided the South Florida respondents into one of six submarkets or regions based on the definition used by the US Department of Housing and Urban Development (HUD) in its analyses of the conditions for growth or decline in housing demand (Kane 2019). We focused on six market areas where we had at least 50 responses: Tampa-St. Petersburg-Clearwater, Cape Coral-Fort Myers, Fort Lauderdale-Pompano Beach-Deerfield Beach, Miami-Dade, West Palm Beach-Boca Raton-Delray Beach, and North Port-Sarasota-Bradenton.

We found that respondent demographic characteristics varied among the different housing submarkets. For example, Realtors® in Miami-Dade had fewer years in the profession, with only 14% reporting having been in real estate for more than 20 years, in contrast to West Palm Beach-Boca Raton-Delray Beach where 40% had been practicing real estate for 20 years or more. Miami-Dade real estate agents were also younger, with 38% of them being over age 55, in contrast with Fort Myers-Cape Coral, West Palm and North Port-Sarasota where two-thirds reported being over age 55. In four of the submarkets, most respondents were male, but in Tampa-St Petersburg-Clearwater and North Port-Sarasota-Bradenton the majority was female. Most respondents in five of the submarkets classified themselves as "white" but in Miami-Dade only 37.5% were "white" and 47.5% identified themselves as "Hispanic."

The political affiliations of the respondents also varied by submarket. In the three Gulf submarkets of Tampa-St. Petersburg-Clearwater, Cape Coral-Ft. Myers, and Northport-Sarasota-Bradenton, more than half of the respondents identified themselves as Republican (Fig. 6.2). In contrast, in Miami-Dade nearly 50% reported that they were political Independents. The highest proportional representation of Democrats was in West Palm Beach-Boca Raton-Del Ray Beach, as well as Fort Lauderdale, where the political affiliations were nearly equally divided among Republicans, Independents and Democrats.

Respondents were also likely to believe that climate change is taking place and that it is related to sea-level rise. When asked, "to what extent do you believe that the sea-level is rising" more than 70% responded that they believe that the sea-level "probably", "very likely", or "definitely" is rising. Similarly, 70% of the respondents

[2] The US Census Bureau estimated that as of July 1, 2019, 26.4% of Florida's population as of was Hispanic or Latino, including persons of any race.

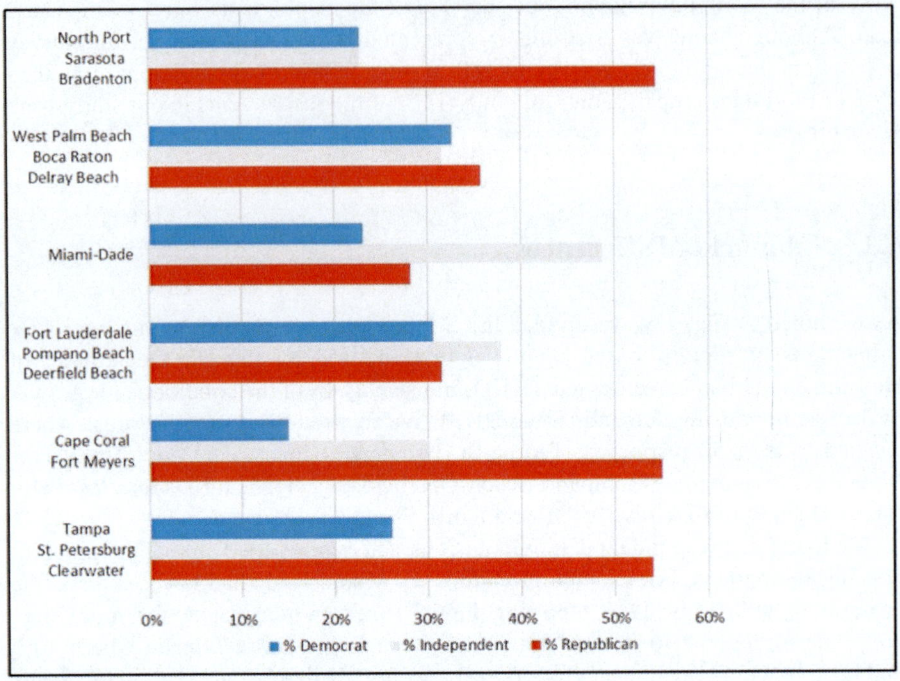

Fig. 6.2 Political identification by submarket

agreed that "climate change is happening." Finally, we asked them "to what extent do you think sea-level rise is related to climate change?" This question elicited a few more "not sure" answers, but nearly 70% agreed that sea-level rise is at least "probably" related to climate change. Despite differences in political orientation as well as ethnic composition by submarket, the beliefs about the reality of climate change, sea-level rise and their interrelationship showed no significant variation by submarket.

6.3 Summary

We collected a relatively large sample of the Realtors® currently selling residential property in south Florida. A total of 330 brokers and 350 sales associates completed the survey, many of whom provided text comments to elaborate on their responses. Of these, nearly 100 provided further elaborations concerning their opinions about the future of the Florida housing market. The real estate agents that participated in this study were generally highly experienced professionals who have lived in Florida for 20 or more years. Most are college graduates. Although the largest percentage of the respondents identified themselves politically as "Independents," more of them at least "lean" towards the Republican Party, with higher concentrations of Republican real estate agents in the Gulf Coast locations. Almost all of them believe that the sea-level is rising, and that sea-level rise is a product of climate change. The survey respondents represent all the important submarkets within south Florida and are well-positioned to provide guidance about trends in the housing market. The next chapters will explore the kinds of information about flood hazards that are now disclosed to prospective homebuyers and how this information has affected the south Florida housing market.

References

Kane KP (2019) HUD's housing market analysis and current state. EMAD housing market methodology. In: HUD User, Office of Policy Development and Research, US Department of Housing and Urban Development. Available at: https://www.huduser.gov/portal/sites/default/files/pdf/emad-methodology.pdf

Linsell C (2020) 67 shocking real estate statistics you need to know. In: The Close. Available at: https://theclose.com/real estate-statistics/

United States Census Bureau (2019) Quick facts: Florida. Available at: https://www.census.gov/quickfacts/FL

Chapter 7
Information About Sea-Level Rise and Its Effects on Home Search Behavior

Ipsa scientia potestas est
Knowledge itself is power
Sir Francis Bacon (1597)

Abstract A great deal of information about property elevation and potential flood risk is available to homebuyers. Most Florida homebuyers are moving from another Florida residence, and therefore are at least somewhat familiar with both storms and tidal flooding. Flood insurance requirements also are a source of information about flood risk. Part of the flood insurance purchase process requires the home buyer to acquire a certificate of elevation that indicates the point at which wave action, tidal action or storm surge will flood the first floor of the house. In 2020, the National Association of Realtors® added two more indicators of flood risk to their realtor.com listings: information about the FEMA insurance classification of the property as well as maps and text from Flood Factor presenting a numerical score representing the flood risk factor of the property on a scale from 1 to 10. At the time of our survey, two to three months after this change, Realtors® reported little response to this additional information from either buyers or sellers. The respondents to the survey indicated that some homebuyers are avoiding low-lying property susceptible to flooding, but most real estate agents stated that it is not the flood risk itself but instead the cost of flood insurance that drives buyer response to coastal property.

Keywords Flood insurance · FEMA · Flood risk

Historically, major property development in Florida was based on the principle of *caveat emptor* or "buyer beware." As described in Chap. 1, there were two Florida land booms in the twentieth century: the first from 1920–1925, and the second in the 1950s and 60s, based on new technology such as air conditioning and the use of the now-banned pesticide DDT for mosquito control, as well as increased personal wealth and leisure time. The more recent era was notable for the use innovative high-pressure sales techniques, including property giveaways on television game shows. As Mormino (2005) describes the sale of undevelopable property on the installment plan, "the future belonged to an eclectic collection of hucksters and

© The Author(s), under exclusive license to Springer Nature Switzerland AG 2022
R. Palm, T. Bolsen, *Housing Market Response to Sea-Level Rise in Florida*, Coastal Research Library 37, https://doi.org/10.1007/978-3-030-88435-2_7

boosters, risk takers and fortune makers who saw retirement communities and instant cities where others saw sand and swamp".

In recent years, real estate disclosure laws have become stricter and there has been greater media attention to the risks of sea-level rise and tidal flooding to coastal communities (Pino 2006). With new regulations and the greater availability of information about environmental risk, it would be reasonable to expect that buyers would be more likely to evaluate flood risk as part of their search for a new home. Indeed, numerous empirical studies have documented the necessary (but often not sufficient) relationship between risk perception, personal preparedness, and hazard mitigation (Harvatt et al. 2011; Mileti et al. 2004; Bodoque et al. 2019; Tierney 2019; Lall and Deichmann 2010; Saarinen 1982; Wachinger et al. 2013).

In this chapter, we will explore the nature of information that is available to prospective Florida homebuyers, and what real estate agents report with respect to whether prospective homebuyers are avoiding purchases in low-lying coastal areas. Both are key to understanding the perspective of real estate agents on the question of whether susceptibility to sea-level rise is affecting house prices in Florida.

7.1 Information Available to Homebuyers About Flood Risk

Several studies of the responses of buyers to areas at risk of flooding have suggested that with more or better information, people will make more well-informed decisions, either choosing to avoid flood-prone areas or exacting price discounts. At present, Florida homebuyers have several sources of information about the flood-risks associated with any property. These include (1) prior familiarity with the area, (2) information freely available on government websites, (3) information obtained in the process of buying flood insurance, and (4) information from the real estate industry itself.

7.1.1 Familiarity with the Area: Most Homebuyers Are Local

About 10% of Americans moved to a different house in 2018. Of these movers, 85% moved within the same state. Although the places of origin for may vary, most Florida homebuyers are moving only short distances. As an example, let's contrast the origins of homebuyers in two Florida metropolitan areas, Fort Myers-Cape Coral on the Gulf Coast and Miami on the Atlantic coast. (US Census Bureau 2018). In 2016, 17.5% of the population of Fort Myers-Cape Coral had changed residence, with most of these people moving from another place within the same county or from Miami-Dade. Naples or Tampa-St. Petersburg. Miami-Dade had a smaller fraction of movers (11.5% in 2016) with most (75%) having moved from the same county. In short, most Florida homebuyers already live in the state, and will have had a great deal of exposure to the idea that sea-level rise is threatening the coast from the popular press, among other sources.

7.1.2 Information from Federal and Local Government

Prospective buyers can view and download maps of the various flood hazard areas used by the Federal Emergency Management Agency (FEMA) for flood insurance classification. These maps are based on a combination of canal and stream flows, tides associated with storms, rainfall, hydraulic analysis, and topographic surveys, and are updated periodically. The FEMA website as well as those of county governments permit detailed searches of public information about the property for sale. For example, the Miami-Dade County home page (miamidade.gov) has an icon on "property search" where, after inputting either an address or a location on a detailed map, property information including the owner's name, lot size, living area, sales and assessment information, and zoning information are provided. Although this link does not provide flood information, one can enter the term "flood" and be taken to a page where flood zones for the entire county as well as for individual parcels are displayed on an interactive map. Similarly, the Pinellas County homepage has an icon for "maps/aerials/property" where the prospective buyer can find information on the FEMA zoning, as well as their emergency evacuation zone and a map of storm surge levels with an illustration of the projected flooding (Fig. 7.1).

Maps of projected flooding are also available from independent organizations such as Climate Central, which has produced "surging seas" risk maps for US coastal areas. The maps available on the Climate Central website portray flooding under different assumptions of the amount of sea-level rise, along with maps of the demographic characteristics of the current residents and estimates of property values per acre (https://sea-level.climatecentral.org).

7.1.3 Elevation Certificate in the Process of Purchasing Flood Insurance

The Federal Emergency Management Agency (FEMA) classifies property as within a Special Flood Hazard Zone if it has at least a 1% chance of being flooded in any given year, or a 26% chance of flooding over the life of a 30-year mortgage. Buyers of property located within a Special Flood Hazard Zone are required to purchase flood insurance. Maps indicating the boundaries of the Special Flood Hazard Zone are widely available from FEMA itself or from the county, city or town in which the property is located, and are often in the format of an "interactive map" in which the

Fig. 7.1 Pinellas County storm surge. (Source: Pinellascounty.org) The depth of floodwaters for the house indicated with the red dot is shown in the picture above, along with information about evacuation from that location

user can enter the street address of a specific property and get local flood information.

Most home purchasers apply for a mortgage loan, and most mortgage loans are made by institutions insured by the federal government. For property in a designated flood hazard zone, such lenders must require the purchase of flood insurance, and the buyer must provide proof of such insurance to complete the home purchase. The insurance policy itself is issued by a private insurance company. Before an insurance policy is issued, the insurer calculates the premiums for the policy based partly on an "elevation certificate". This documentation is mandatory if (1) the mortgage loan is financed by a federally regulated or insured lender, (2) the property is in a Special Flood Hazard Area, or (3) the policy itself is backed by the NFIP. Elevation certificates for specific properties may already exist and be available from the seller

U.S. DEPARTMENT OF HOMELAND SECURITY
Federal Emergency Management Agency
National Flood Insurance Program

OMB No. 1660-0008
Expiration Date: November 30, 2018

ELEVATION CERTIFICATE

Important: Follow the instructions on pages 1–9.

Copy all pages of this Elevation Certificate and all attachments for (1) community official, (2) insurance agent/company, and (3) building owner.

SECTION A – PROPERTY INFORMATION | FOR INSURANCE COMPANY USE

A1. Building Owner's Name | Policy Number:

A2. Building Street Address (including Apt., Unit, Suite, and/or Bldg. No.) or P.O. Route and Box No. | Company NAIC Number:

City State ZIP Code

A3. Property Description (Lot and Block Numbers, Tax Parcel Number, Legal Description, etc.)

A4. Building Use (e.g., Residential, Non-Residential, Addition, Accessory, etc.) ______

A5. Latitude/Longitude: Lat. ______ Long. ______ Horizontal Datum: ○ NAD 1927 ○ NAD 1983

A6. Attach at least 2 photographs of the building if the Certificate is being used to obtain flood insurance.

A7. Building Diagram Number ______

A8. For a building with a crawlspace or enclosure(s):

a) Square footage of crawlspace or enclosure(s) ______ sq ft

b) Number of permanent flood openings in the crawlspace or enclosure(s) within 1.0 foot above adjacent grade ______

c) Total net area of flood openings in A8.b ______ sq in

d) Engineered flood openings? ○ Yes ○ No

A9. For a building with an attached garage:

a) Square footage of attached garage ______ sq ft

b) Number of permanent flood openings in the attached garage within 1.0 foot above adjacent grade ______

c) Total net area of flood openings in A9.b ______ sq in

d) Engineered flood openings? ○ Yes ○ No

SECTION B – FLOOD INSURANCE RATE MAP (FIRM) INFORMATION

B1. NFIP Community Name & Community Number	B2. County Name	B3. State

B4. Map/Panel Number	B5. Suffix	B6. FIRM Index Date	B7. FIRM Panel Effective/ Revised Date	B8. Flood Zone(s)	B9. Base Flood Elevation(s) (Zone AO, use Base Flood Depth)

B10. Indicate the source of the Base Flood Elevation (BFE) data or base flood depth entered in Item B9:
○ FIS Profile ○ FIRM ○ Community Determined ○ Other/Source: ______

B11. Indicate elevation datum used for BFE in Item B9: ○ NGVD 1929 ○ NAVD 1988 ○ Other/Source: ______

B12. Is the building located in a Coastal Barrier Resources System (CBRS) area or Otherwise Protected Area (OPA)? ○ Yes ○ No

Designation Date: ______ ○ CBRS ○ OPA

FEMA Form 086-0-33 (7/15) Replaces all previous editions. Form Page 1 of 6

Fig. 7.2 Elevation certificate required for Flood Insurance. (Source: FEMA.gov)

or a floodplain manager, but otherwise the buyer can obtain an elevation certificate by hiring a professional surveyor. Information in this certificate includes the elevation of the bottom floor or basement, the garage, and the machinery or equipment that services the building such as the heating system (Fig. 7.2).

7.1.4 *Information from the Real Estate Industry*

The real estate industry also has undertaken independent flood hazard disclosure. Before the National Association of Realtors® changed their website in the summer of 2020, a few messages appeared with property listings on Zillow in 2018. These directed the viewer to a website supported by the First Street Foundation where they could learn more in response to a warning like this: "Flooding Hurts Home Value. Know before you buy. Find out for free now. FloodIQ.com." By clicking on this link, one would be taken to a description of the flood risk associated with sea-level rise projected for the site in the next 15 years, as well as a visual portrayal of flood scenarios under conditions of high tide and hurricanes of various intensities (Fig. 7.3).

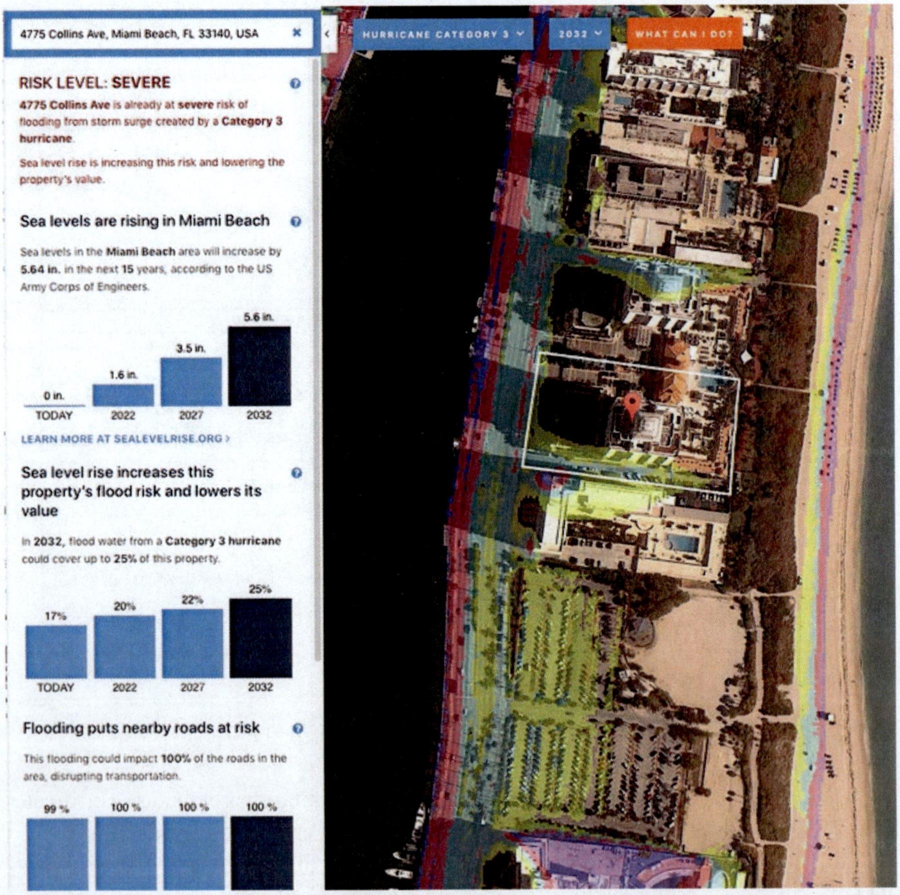

Fig. 7.3 Detail from FloodIQ.com link in 2018. (Source: Flood risk data is provided by Flood Factor®, a product of First Street Foundation®)

This linkage between property advertisements and hazards warnings was an innovative step in providing very specific hazards information to prospective buyers. Although Zillow removed these notifications from their listings, the First Street Foundation reached an agreement in August 2020 with the National Association of Realtors® to include their maps, now known as "Flood Factor," on all of their listings (First Street Foundation 2021). An NPR report on this agreement asserted that this move would be "a major shift in consumers' access to information about climate threats" (Hersher and Sommer 2020). The reporter noted that other housing websites such as Redfin, Zillow, and Trulia "have no plans to share information about flooding with users" because they say that "home sellers are reluctant to publish flood risk information, since it could decrease their home's value" (Hersher and Sommer 2020). Redfin shifted their position on this issue in February 2021 when they announced that they would begin to put the "flood factor" score as well as the FEMA insurance zone on their property listings (Dickerson 2021). The Redfin presentation of this information is less prominent than that in realtor.com, displaying the flood information only after thirteen other items: a photo and general description, "price insights", "home facts", map, open houses, payment calculator, nearby similar homes, property details, "ask a question" featuring the selling agent, sales and tax history, schools, "activity" including views and tours, and "public facts". The chief product officer at Redfin justified including "flood factor" as making it possible for people to "make better-informed decisions about buying and selling" as they integrate "the frequency or intensity of natural disasters ... into their decision about where and whether to buy or sell a home" (Redfin Journalist Services 2021).

The NAR decided to include two sources of information on flood risk: the FEMA flood zone and the "Flood Factor" rating. They also include a statement about whether flood insurance is required, with advertisements for private companies offering to give an estimate of the annual premium for flood insurance[1] Clearly, the expectation of the First Street Foundation, Redfin, and the National Association of Realtors® (NAR) is that the new website information will have a major impact on buyer awareness of flood risk.

[1]In early 2021, two companies were advertised. Tryflood, LLC is a limited liability company incorporated in Florida in June of 2020. Its directors include Chris Walker, the Corporate Service Company and Wright National Flood Insurance Services. Neptune Flood Insurance was created in 2016 and is also headquartered in Florida. It offers insurance outside the NFIP, using what it says is "the most accurate and comprehensive elevation data in the world to make sure you aren't overcharged for flood insurance," and allowing the buyer to purchase flood insurance "in two minutes or less."

7.2 Response of Buyers and Sellers

The website was launched on August 26, 2020, and the survey of real estate brokers and sales associates was conducted between October 9 – November 23, 2020. This means the site was relatively new (6–12 weeks old) when the real estate agents we surveyed were reacting to it. We asked whether the respondents had "seen any response to this new information from buyers or sellers". Most reported no response from either. Only 9% of the respondents said that there had been a response to this information from homebuyers and only 5% reported a response from sellers, with response options that included "yes," "no" and "not sure" for both groups. We followed up with these 62 respondents asking whether the response from buyers and/or sellers had been positive or negative (on a 7-point scale). The response they reported was mixed: most reported a positive or neutral response from buyers. Seller reaction was more likely to be neutral or negative.

7.3 Realtor® Support for the New Website

We next asked all respondents, "What is your view of the decision of the National Association of Realtors® to add this "Flood Factor" information to property listings?" (1= "strongly oppose", 4= "neither support nor oppose", 7= "strongly support"); and "How likely are you to recommend that potential buyers view this new "Flood Factor" information included on property listings?" (1= "extremely unlikely", 4= "neither likely nor unlikely", 7= "extremely likely"). Most real estate agents we surveyed (54%) expressed some level of support (i.e., a score higher than "4" on the 7-point response scale) for the NAR's decision to add the "Flood Factor" information to the property listing (Table 7.1). While one-third of respondents selected that they "neither supported nor opposed" the decision, only 13% expressed any level of opposition to it. Most respondents also said that it was "likely" that they would recommend this information to potential buyers (52%), while only 18% said that it was "unlikely" that they would do so.

We included an open-ended response item on the survey that asked all respondents, "In a few sentences, please give us your opinion about the addition to the

Table 7.1 Realtor® support for NAR's decision and likelihood will recommend to others

	Support	**Oppose**	**Not sure**
View on NAR decision	54% (370)	13% (89)	33% (221)
	Will Recommend	**Will Not Recommend**	**Not sure**
Will you recommend that buyers consult it?	52% (352)	18% (125)	30% (203)

listings of these maps and the information that accompanies them" to which 587 responded. Respondent remarks were largely positive and reflected the importance of full disclosure and transparency as a guiding ethic of their work. One respondent stated, "*I'm all for it, as it better informs potential buyers of the risks associated with the location and should trigger the pursuit of a flood insurance estimate where needed so they will be aware of the 'carrying cost' associated with the flood zone.*" Similarly, another respondent offered, "*As brokers and agents, transparency is key and crucial to our business. It is important that the buyer makes a fully informed decision when purchasing a home.*"

Many respondents commented on the importance of providing as much information as possible to potential buyers: for instance, "*I believe the more information you disclose about a property the better it is for all parties involved.*" Another added, "*It is the Realtor's responsibility to "FULLY INFORM" every Buyer about existing and potential problems of every property they represent.*" In addition, many expressed a positive view about the importance of this information in helping buyers make an informed decision. One wrote: "*Buyers NEED this information. Realtors, unfortunately, fail to provide information about flood areas – they direct potential buyers to do their 'due diligence.' This posted information is a start.*"

A small percentage expressed opposition to the NAR's decision to include the Flood Factor information. For example, one said: "*Adding these only adds unnecessary complications to a transaction. Areas that were not in flood zones now are, even though they may have not had any issues with flooding.*" Another respondent expressed reservations about the accuracy of the Flood Factor maps: "*This information presents a very poor graphic as to the true flood factor. Flood zones present a risk factor based on a 50- or 100-year flood, not at any single point in time... and these floods may not happen for years.*" Another recalled his or her own experience with the area: "*I have sold property in that exact neighborhood and have never seen flooding there.*"

Further, some expressed concerns about the effect of this information on property sales and sellers: "*It may scare some buyers away from properties that may never experience a flood.*" Another stated, "*all it will do is cause property values to go down and scare people and probably make some properties unsellable – I guess that might be what their intention is – scare and stop sales.*". More colorfully, another respondent stated, "*This sounds overblown and biased to panic new low-end buyers. Flooding is about as concerning to many of us as being abducted by aliens... This is very concerning to 1 out of every trillion or so people...*"

7.4 Do Prospective Buyers Avoid Seeking Property in Low-Lying Coastal Areas?

From the variety and quality of information sources just reviewed, it seems clear that prospective homebuyers have a rich array of facts about flood risk available to them, some of which are highly specific about the individual property being considered.

Further, most Florida homebuyers are moving from a previous residence within Florida, and therefore have had experience with the climate and climate risks of Florida. If people are purchasing property at risk of flooding, it should not be because information about the risk is not available to them.

One of the basic questions we needed to answer about the operation of the Florida housing market was buyer response to property at risk of flooding, particularly given sea-level rise. If some segment of the population is avoiding property at risk of flooding, then demand should decline, and prices should fall. We therefore asked the real estate agents whether they observe that buyers are avoiding properties susceptible to flooding. The question was worded: "To what extent do you believe that buyers are avoiding property in low-elevation coastal areas because of possible flooding?" (Fig. 7.4). Overall, real estate agents were divided in their observations of whether buyers avoid low elevation property. A little under a quarter of the respondents (23%) reported that buyers rarely or never avoid such property. Almost half selected the "middle" option of "somewhat" and thirty percent said "somewhat" or "somewhat frequently". The mean response was 3.78 (closer to "somewhat" than "a little" on the 7-point scale from "not at all" to "all the time").

The aggregate responses mask the important submarket differences in the distribution of responses, however (Table 7.2). The profile of responses of was highly varied with modal responses of "rarely avoid" in West Palm-Boca-Del Ray and Miami-Dade, "a little" in Northport-Sarasota, and "somewhat avoid" in Tampa-St. Petersburg, Cape Coral-Fort Myers, and Fort Lauderdale.

Comments from the respondents revealed something of the reasoning behind these responses. Those who responded "frequently" usually added something about flood insurance costs. For example, one stated that: "*Purchasing a mandatory flood insurance policy will increase out of pocket expenses and reduce spendable income*," or similarly, "*They are mostly concerned due to the cost of the flood*

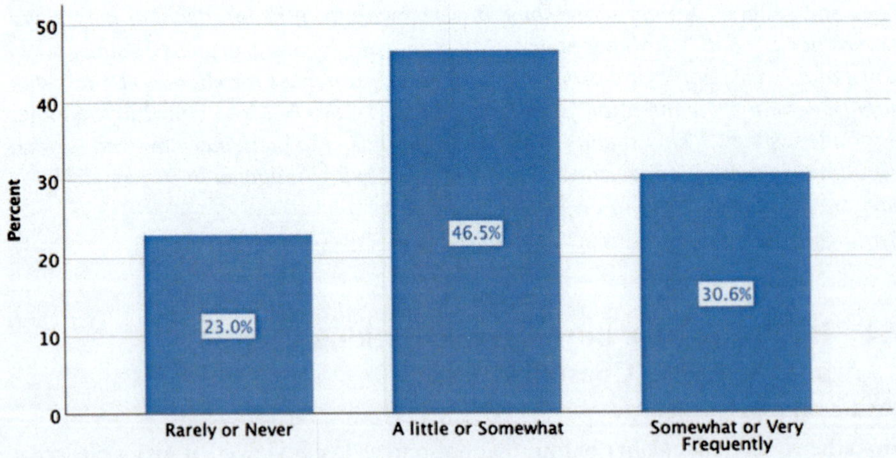

Fig. 7.4 Are buyers avoiding low-elevation coastal property?

Table 7.2 Submarket responses to avoiding low-elevation coastal property

	Tampa(%)	Ft My-ers(%)	Ft Laud.(%)	Miami Dade(%)	West Palm-(%)	North port(%)	Total(%)
Not at all	1	6	5	5	6	2	4
Rarely	12	16	19	27	31	17	20
A little	23	19	20	19	19	31	21
Somewhat	25	27	34	25	25	15	26
Somewhat frequently	19	19	16	9	11	12	15
Very frequently	19	8	8	15	8	20	13
All the time	1	5	0	1	1	3	2

insurance as it is mandated for financed purchase, and not necessarily as concerned with the actual threat of flooding". Several noted that the avoidance was more common for lower income buyers: "*Buyers with less secure income absolutely do NOT want to live in a flood zone. Buyers with a lot of diversified assets do not mind the risk and prefer to live right on the water.*" Another compared first-time buyers with those with more resources: "*Most first-time homebuyers don't have the extra funds for flood insurance. 2nd or 3rd time homebuyers may be ok moving in a flood zone*". Others described it as a trade-off between a desirable location on the coast and the cost of protecting that location: "*Buyers are concerned with flooding, primarily because of the cost of flood insurance. Most properties I sell are in a flood area, so it comes down to elevation and premium cost. Some of the most desirable neighborhoods are flood zones.*" A few, very few, respondents noted that coastal areas are becoming more at risk with sea-level rise and that their clients are responding to these changes: "*Sea-level is rising faster than originally thought*" and "*climate change has made designated flood areas important*".

Those who answered that they had rarely if ever had a client mention elevation or flood risk to them tended to note the high demand for coastal property or the general lack of concern for flood risk when compared with the amenities of a beach location. For example, "*I've never had a single client bring up the issue. No one wants their property to flood but no one buying here is concerned about that actually happening.*" Another common response referred to the demand for beach property: "*Those are the best neighborhoods, and most desirable locations*" or "*Most buyers are not deterred by low lying coastal areas and are happy to pay flood insurance premiums in exchange for being close to the beach.*" Some noted that homebuyers are not concerned with flooding: "*they are either ignoring the risk or not cognizant of the risk*" or "*they think global warming and rise of sea-level is in the future and will not affect them now.*" A few of the respondents included political opinions: "*we are in Trump country, and he believes there is no climate change*" and "*the flooding issue is an ideological argument without any fact.*"

7.5 Do Investors Differ from Owner-Occupiers?

Previous research (Bernstein et al. 2019) noted that investors or non-owner-occupiers are likely to be more aware of the risks of sea-level rise and therefore extract greater price discounts. In a detailed comparison of the impacts of elevation on price discounts for Pinellas (St. Petersburg-Clearwater) as opposed to Miami-Dade County, Fu and Nijman (2021) found a complex pattern of relationships: for all buyers in Pinellas County, they found discounts to property at elevations below 3 feet, but in Miami-Dade, no price impact on property below 2 feet and instead discounts for property at 2–4 feet in elevation. Nonprimary homeowners or second homebuyers in Miami-Dade County paid relatively more than primary homeowners at elevations of less than 2 feet but primary homeowners paid more than nonprimary homeowners in Pinellas County for property at 3–4 feet in elevation. Fu and Nijman attributed these market differences to several factors, including higher average elevations in Pinellas County and different buyer types, specifically the insensitivity of very high-income buyers to properties at risk from sea-level rise among the affluent Miami-Dade buyers investing in second homes. We wanted to see if real estate agents had observed such differences and posed this question: "*We might expect that buyers who are planning to live in the property year-round themselves might respond differently to possible flooding than those investing in property or buying it as a vacation home. In your opinion, do different types of buyers avoid properties in low-lying coastal areas*?" We asked them to respond on the same 7-point scale from "not at all" to "always."

Interestingly, when asked to consider whether specific types of buyers are avoiding vulnerable property, there was more of a tendency to respond that buyers of all types are not avoiding low elevation property. Although the mean score for whether buyers generally are avoiding these areas was 3.78, the mean scores for owner-occupier, second-homebuyers and investors were consistently lower, at 3.31, 2.89 and 2.97 respectively, indicating an average answer of "a little" or "rarely", reflecting a belief that these classes of buyers were less likely to avoid such property. Overall, the real estate agents reported that prospective owner-occupiers were more likely than other buyers to avoid low-lying property, but it should be noted that even for this group, the majority answered "rarely." For investors and second

Table 7.3 Investors, second homebuyers, and owner-occupiers with respect to avoiding low elevation coastal property

Investors			Second home		Owner	
	Frequency	Cumulative %	Frequency	Cumulative %	Frequency	Cumulative %
Rarely	421	65	442	66.7	374	55.2
Somewhat	126	84.4	106	82.7	139	75.7
Frequently	101	100	115	100	165	100

Table 7.4 Avoiding low elevation property by submarket and property use by buyer

	Owners	Investors	Second home	Mean
Tampa-St. Petersburg-Clearwater	3.53	3.02	2.91	3.15
Cape Coral – Fort Myers	3.3	3.16	3.23	3.23
North Port- Sarasota-Bradenton	3.47	3.38	3.34	3.4
West Palm Beach-Boca Raton-Delray	3.04	2.59	2.76	2.8
Ft Lauderdale-Pompano-Deerfield	3.11	2.69	2.79	2.86
Miami-Dade	3.3	2.46	2.84	2.87

homebuyers, two-thirds responded that clients rarely avoid such property (Table 7.3).

We found only small submarket differences in these opinions about the search patterns of different types of buyers (Table 7.4). Mean scores on the 1–7 scale clustered around 3.0 or "a little." The two submarkets where the real estate agents reported the greatest divergence among buyer types were Miami-Dade and Tampa-St. Petersburg-Clearwater. In both cases, the real estate agents observed prospective owner-occupiers were more likely to avoid low-lying coastal areas than investors. When asked to elaborate on why they believed these differences exist, the respondents tended to mention that owner-occupiers would be more concerned with living in the property and therefore would be more sensitive to flood risk. For example, one said: "*owner occupiers care more because they will actually have to deal with the flooding themselves. Investors and second homebuyers are less likely to be present when flooding occurs*" or "*owner-occupiers for the most part consider the 'livability' of the properties and while second homeowners/investment buyers are looking for income potential.*"

Respondents repeatedly compared the greater willingness of wealthier homebuyers to self-insure or simply to take on the risk of flooding: "*most investors or second homebuyers have better income/savings and can afford additional insurance costs*" or "*the difference is income and financial resources. If a buyer can afford to purchase on the beach and that is their driving motivation, there is normally NO concern about flooding … it's the potential risk of living on the beach.*" Several respondents noted that "*there is always a buyer willing to take a risk to be in a prestige location for the right price*" or "*they are willing to roll the dice as long as the flood insurance is low enough.*"

The two submarket areas where the real estate agents saw little difference among the three groups of buyers were Cape Coral-Fort Myers and North Port-Sarasota-Bradenton, both on the Gulf Coast. Here the real estate agents attested that their buyer type mattered little with respect to low-lying coastal areas: "*flood risk is something buyers don't consider here typically*" or "*high net worth individuals seem less concerned about natural disasters*" or "*cash buyers in any category are not forced to buy flood insurance and therefore don't consider it a major factor in deciding where to buy.*" One asserted that: "*this is a largely Republican state, and they don't believe in global warming. Therefore, I don't think they believe sea-level rise is a reality*"

7.6 Do Buyers of Less Expensive Houses Avoid Low Elevation Property More Frequently?

We also expected that lower income buyers might differ from higher income buyers with respect to avoiding coastal property. Since the cost of flood insurance would be more of a burden to households on a tight budget, we expected that buyers of less expensive houses would be more focused on avoiding low lying property to save on insurance costs. We posed the question as: "*Does elevation matter more or less for homes that are at the high end of the market as opposed to those selling for lower prices?*" Respondents could choose from options stating that elevation matters more at the high end, at the low end or that there are no differences. We found, counter to our expectation, that 42% of the respondents felt there was no difference between low-end or high-end buyers with respect to concern with elevation (n=289), while 32% reported that it matters more at the *high end* of the market (n=217), and 26% (n=172) said that it matters more at the low end. We can conclude from this distribution of responses that real estate agents are not observing systematic differences in response to elevation based on the cost of housing.

7.7 Real Estate Agent Characteristics and Opinions About Buyer Behavior and Flood Factor

The professional experience of real estate agents, their demographics, and their general beliefs about sea-level rise and climate change may impact the way they interpret the behavior of their clients with respect to avoiding coastal properties. These variables may also help us understand differences in opinions about the NAR's decision to add Flood Factor information available to property listings. If real estate agents are worried about sea-level rise and the prospect of increased future flooding, they may convey this attitude to the buyers they work with. They could interpret buyer preferences through their own knowledge sets and beliefs about sea-level rise. Alternatively, if real estate agents do not believe that climate change and sea-level rise are a real threat, they may dismiss the notion that their buyers are concerned with low-elevation property. Of course, we cannot discern from the questions included on our survey whether agents are conveying their own beliefs to prospective buyers or interpreting buyer preferences based on their own beliefs. However, in this section, we examine the degree to which specific characteristics of real estate agents are associated with variance in respondents' views about the propensity of homebuyers to avoid property at low elevations susceptible to flooding, and their support for Flood Factor as a tool for informing potential buyers.

First, we might expect that the perceptions of real estate agents with respect to the impacts of sea-level rise on the housing market might be affected by their own beliefs about the reality and effects of climate change. We asked if they agreed that the sea-level is rising and that climate change is occurring, and whether they felt sea-level rise and climate change were related. As noted earlier, responses to these three questions were highly inter-correlated (Cronbach's alpha = 0.94), and therefore we created a single scaled variable to evaluate real estate agents' views about climate change induced sea-level rise (*Sea-level Scale*).

Second, we might expect that the same demographic and attitudinal variables that have been found to predict beliefs about sea-level rise in the general population would also predict real estate agents' perceptions concerning the propensity of buyers to avoid flood-prone areas, as well as their own opinions about the NAR's decision on including Flood Factor. In supplementary analyses reported in the Appendix 3, we evaluated the extent to which several key political and demographic characteristics of our respondents – political party identification, age, level of education, gender, ethnic identity, and location – are associated with variance in responses for the *Sea-level Scale*. We note that, as many previous surveys have found, the party identification of real estate agents is the strongest predictor of variation in real estate agents' beliefs about climate change induced sea-level rise. Higher levels of education were also associated with more belief in climate-induced sea-level rise. We investigated the association of these variables with perceived buyer avoidance of low-lying property.

Third, we would also expect that real estate agents have different degrees of familiarity with the housing market which could affect their perceptions of buyer attitudes. Agents with many years of experience in real estate and great familiarity with the local area in which they are selling houses may be able to convey a confidence to buyers that lets them minimize the risk of flooding, and as a result, regardless of the actual views of the clients, the agents may believe that the buyers too are not concerned with flood risk. Two measures of real estate agent experience with the market we included were: (1) the level of activity they have in their real estate practice as measured by the number of houses they sell per year (*Houses Sold*) and (2) the number of years they have been in the real estate profession (*Years in FL*).

Finally, we might expect that market location on the Gulf or on the Atlantic coast could affect perceived buyer preferences for coastal property. As we have noted earlier, homebuyers on the Atlantic coast are more likely to have larger representation from investors as opposed to prospective owner-occupiers. Since owner-occupiers have a deeper connection to the Florida community to which they are moving, they are also more likely to convey concern about living at low elevations susceptible to flooding.

Using three ordinary-least squares (OLS) models, we evaluated the factors associated with differences in respondents' views across the 7-point response scales for three items. The first is labelled "*avoiding*" and represents the question to what extent are buyers avoiding low elevation coastal property. The second, labelled "*decision*" is the response to the question about opinions concerning the decision of the NAR to add "Flood Factor" to property listings. The third, "*recommend*", is the response to how likely the real estate agents say they are to recommend this new information to potential buyers (Table 7.5).

Several factors were associated with real estate agent views about the degree to which buyers are avoiding low-elevation properties and their response to the NAR information on the realtor.com website. First, the "sea-level scale" (climate change beliefs) was a strong predictor of all three items: buyer avoidance of low-lying property, support for the NAR decision to include Flood Factor in the listing, and likelihood of recommending this information to clients. Older real estate agents were also more likely to report buyer avoidance of low-lying property and were also more likely to say they would recommend the flood risk information. Real estate agents on the Gulf Coast as opposed to the Atlantic Coast were more likely to report buyer avoidance of low-lying property. In addition, there was a negative association between the number of houses an agent sells and both their support for the NAR's decision and the likelihood that they will recommend the information. Finally, female real estate agents were more likely to say that they would recommend the new information to prospective buyers. Other demographic characteristics appear to have accounted for little variance in respondents' views toward the NAR's decision.

7.8 Summary

Prospective homebuyers have a wide range of information sources from which they can learn about property elevation and potential flooding. Since most Florida homebuyers are moving from within the state, we can assume they are at least somewhat familiar with hurricanes, storm surge and tidal flooding, if not from personal experience, then certainly from media reports. For those moving from out-of-state, there are a variety of information resources available from the federal and county governments as well as privately-run websites. Further, as of August 2020, realtor.com itself has provided information both from the First Street Foundation and FEMA as to likely flooding.

When real estate agents were asked to think about whether purchasers who are going to live in the property, buy it as an investment, or buy it as a second home are avoiding areas in low-lying coastal areas, their average answer was "a little." They observed more prospective owner-occupiers than investors bypassing such

Table 7.5 Real estate agents' Characteristics and Beliefs about Buyer Behavior and Flood Factor

	(1) Avoiding		(2) Decision		(3) Recommend	
	Coef.	*p-value*	*Coef.*	*p-value*	*Coef.*	*p-value*
Age	.10*	0.05	0.06	0.32	0.25***	0.00
	(0.05)		(0.05)		(0.07)	
Education	–.08	0.07	0.05	0.31	–.06	0.30
	(0.04)		(0.05)		(0.06)	
Female	.18	0.12	0.14	0.22	.33*	0.03
	(0.12)		(0.12)		(0.15)	
Minority	.25	0.09	0.03	0.95	.36	0.06
	(0.15)		(0.15)		(0.20)	
Republican	.15	0.28	–0.12	0.40	–.07	0.70
	(0.14)		(0.15)		(0.18)	
Democrat	–.02	0.94	–0.16	0.32	–.15	0.47
	(0.16)		(0.16)		(0.21)	
Sea-level scale	.15***	0.00	0.32***	0.00	.21***	0.00
	(0.05)		(0.05)		(.06)	
Years in FL	.02	0.772	–.03	0.60	.08	0.25
	(0.06)		(0.06)		(0.07)	
Houses sold	–.17	0.191	–0.35**	0.01	–.35*	0.03
	(0.13)		(0.13)		(0.17)	
Atlantic	–.47***	0.00	–0.06	0.64	–.13	0.43
	(0.12)		(0.13)		(0.16)	
Constant	2.87***	0.00	2.98***	0.00	2.27***	0.001
	(0.54)		(0.55)		(0.70)	
Observations	614		612		612	
Adjusted R^2	0.043		0.102		0.051	

*$p < 0.05$, **$p < 0.01$, ***$p < 0.001$

properties. Overall, however, large numbers of respondents report that buyers rarely or never take either elevation or likely flooding into account when searching for a home. Real estate agents whose market area was the Gulf Coast were more likely than those on the Atlantic coast to have noticed buyers avoiding low-lying property. Few reported having seen a response to the new realtor.com website from either sellers or buyers. Finally, there is an association between the observations of the real estate agents and their own beliefs about climate change and sea-level rise: those who were more likely to agree that the sea-level is rising, that climate change is happening and that these two are related are also more likely to report observing clients avoiding low-elevation coastal areas susceptible to flooding.

In sum, although there are some real estate agents who report never having had a client shy away from low-lying coastal property, the majority have had at least a few clients mention this consideration to them. If this behavior is systematic and widespread, the market should see evidence of it in the form of weakened overall demand and relatively lower house prices. That is the topic we will explore in the next chapter.

References

Bernstein A, Gustafson MT, Lewis R (2019) Disaster on the horizon: the price effect of sea-level rise. J Financ Econ 134(2):253–272. https://doi.org/10.1016/j.jfineco.2019.03.013

Bodoque JM, Díez-Herrero A, Amérigo M, García JA, Olcina J (2019) Enhancing flash flood risk perception and awareness of mitigation actions through risk communication: a pre-post survey design. J Hydrol 568:769–779. https://doi.org/10.1016/j.jhydrol.2018.11.007

Climate Central (2021) Surging seas. Available at: https://sea-level.climatecentral.org/

Dickerson L (2021) Redfin starts displaying flood risk data on listings. In: Inman newsletter. Available at: https://www.inman.com/2021/02/18/redfin-starts-displaying-flood-risk-data-on-listings/?utm_source=Flood+Factor&utm_campaign=71d02d838c-EMAIL_CAMPAIGN_2020_06_19_02_04_COPY_01&utm_medium=email&utm_term=0_2e3335b90a-71d02d838c-92592982

First Street Foundation (2021) First Street Foundation Mission. Available at: https://firststreet.org/mission/

Fu X, Nijman J (2021) Sea-level rise, homeownership, and residential real estate markets in South Florida. Prof Geogr 73(1):63–71. https://doi.org/10.1080/00330124.2020.1818586

Harvatt J, Petts J, Chilvers J (2011) Understanding householder responses to natural hazards: flooding and sea-level rise comparisons. J Risk Res 14(1):63–83. https://doi.org/10.1080/13669877.2010.503935

Hersher R, Sommer L (2020) Major real estate website now shows flood risk. Should they all? In: National Public Radio, Morning Edition, 26 Aug 2020. Available at: https://www.npr.org/2020/08/26/905551631/major-real estate-website-now-shows-flood-risk-should-they-all

Lall SV, Deichmann U (2010) Density and disasters: economics of urban hazard risk. The World Bank. https://doi.org/10.1596/1813-9450-5161

Miami-Dade County Government (2021). https://www.miamidade.gov/global/home.page

Mileti D, Nathe S, Gori P, Greene M, Lemersal E (2004) Public hazards communication and education: the state of the art. University of Colorado, Natural Hazards Center, Boulder

Mormino GR (2005) Land of sunshine, state of dreams: a social history of modern Florida. University Press of Florida, Gainesville

Pinellas County Government (2021) https://www.pinellascounty.org/

Pino M (2006) Psst ... wanna buy some swampland? In: Orlando Sentinel, 1 Sept 2006. Available at: https://www.orlandosentinel.com/news/os-xpm-2006-09-01-papersubs01-story.htmlrd

Redfin Journalist Services (2021) Redfin publishes flood risk data from Flood Factor for over 94 million homes. Available at: https://press.redfin.com/news-releases/news-release-details/redfin-publishes-flood-risk-data-flood-factor-over-94-million/

Saarinen TE (1982) The relation of hazard awareness to adoption of mitigation measures. In: Saarinen TE (ed) Perspectives on increasing hazard awareness, Program on environment and behavior monograph no. 35. Colorado Institute of Behavioral Science, University of Colorado, Boulder, pp 1–35

Tierney K (2019) Disasters: a sociological approach. Wiley, Hoboken

United States Census Bureau (2018) County-to-county migration flows: 2012–2016. In: American Community Survey. Available at: https://www.census.gov/data/tables/2016/demo/geographic-mobility/county-to-county-migration-2012-2016.html

Wachinger G, Renn O, Begg C, Kuhlicke C (2013) The risk perception paradox-implications for governance and communication of natural hazards. Risk Anal 33(6):1049–1065. https://doi.org/10.1111/j.1539-6924.2012.01942.x

Chapter 8
The Housing Market in Coastal South Florida from the Perspective of the Real Estate Agent

Abstract The survey of real estate agents was intended to provide a perspective from informed professionals on the status and the trajectory of the housing market in coastal south Florida. When asked whether they believed that house prices were relatively stagnant or falling in low-elevation coastal areas subject to flood risks, most responded either rarely or not at all. They also generally have seen little or no evidence that either lenders or appraisers are taking vulnerability to sea-level rise into account when making decisions about home mortgage loans. Most of the real estate agents also foresee little impact of coastal flooding on demand or prices for such properties in the 5–10-year future.

Keywords Real estate appraisers · Home mortgage lenders · Sea-level rise · House prices

A key conclusion of many of the empirical studies of the Florida housing market is homebuyers need more information about flood risk and sea-level rise. They posit that with more information, homebuyers would be better equipped to weigh and assess the impacts of flood risk due to sea-level rise in the purchase process, resulting in a reduction in demand for housing at low elevations along the coast. Lower aggregate demand would be accompanied by stagnating or falling prices, particularly in comparison with properties at higher elevations. Some studies have asserted that this process is taking place right now, and it is therefore possible to observe different price trajectories for coastal property at different elevations. We posited that if it is true buyers are currently taking elevation into account in their search process and purchase decisions, the real estate agents working with them should have taken notice. In this chapter, we review the evidence from our survey on real estate agent views of house price trajectories, interactions with lenders and appraisers concerning home values in flood-prone areas, and the expectations of real estate agents about the 5–10-year prospect for coastal property in Florida.

© The Author(s), under exclusive license to Springer Nature Switzerland AG 2022

R. Palm, T. Bolsen, *Housing Market Response to Sea-Level Rise in Florida*, Coastal Research Library 37, https://doi.org/10.1007/978-3-030-88435-2_8

8.1 Has the Threat of Sea-Level Rise Affected Florida House Prices?

In the previous chapter, we noted that some of the real estate agents responded that some of their clients try to avoid low-lying coastal property, particularly buyers who intend to be full-time residents and are moving to Florida from other parts of the country, buyers who will need mortgage financing and who cannot afford or do not want to pay for the additional costs of mandatory flood insurance, or buyers who simply do not want to worry about flood waters damaging their homes. The question is whether this degree of avoidance has been translated into a reduction in overall demand and therefore a weakening of house prices.

We asked the real estate agents: *"If some buyers are avoiding low-elevation coastal areas, this would reduce overall demand and therefore reduce prices. To what extent have you observed that house prices are either falling or not rising as rapidly for properties at risk of flooding?"* on a 7-point scale from "not at all" (1) to "always" (7).

Responses to this question were quite uniformly negative. Forty-five percent of respondents, the modal category, reported "not at all" (Fig. 8.1). An additional 27% reported that this "rarely" happens, while 22% said it happens "a little" or "somewhat." Numerically, only 11 of 680 respondents indicated that house prices for properties at risk of flooding were "very frequently" stagnating or falling. There was little or no variability among the housing submarkets on this question: the cumulative percentage of responses of the combined categories of "not at all," "rarely" and "a little" ranged from 83% to 87% across the six submarkets.

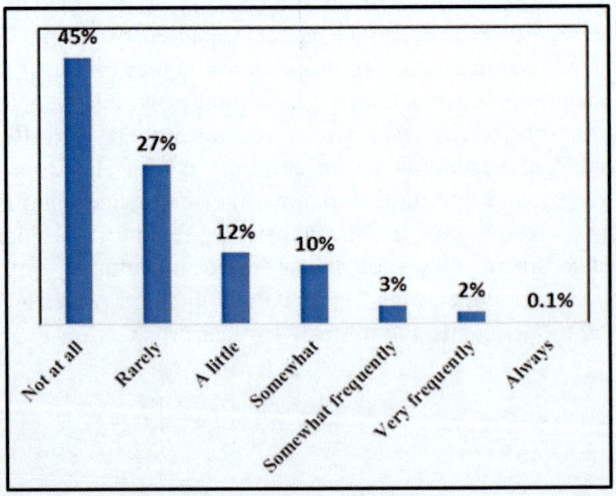

Fig. 8.1 Are house prices for "at risk" properties "falling or not rising as rapidly"?

8.2 Do Lenders Consider Flood-Risk in Their Home Mortgage Decisions?

To probe this notion further, we wanted to learn about their experiences with lenders and appraisers. Homebuyers frequently seek the advice of their real estate agent in deciding on where to apply for a home loan, and there is often a close working relationship between home loan officers and real estate agents (Weintraub 2020). We therefore asked the real estate agents: "*Have you observed mortgage lenders declining loan applications or increasing charges for loans in low-lying areas that are flood prone, in the form of points or mortgage insurance, for example?*" There was even more agreement among the respondents to this question: 60% of the respondents said, "not at all", and only 7% said "somewhat frequently", "very frequently" or "all the time" (Fig. 8.2). There was little variability in this response by submarket: the cumulative percentage of negative response to this question (the sum of those saying, "not at all," "rarely" or "a little") ranged from 82% in West Palm-Boca-Del Ray to 90% in Tampa-St. Petersburg-Clearwater.

Since mortgage loans are commonly made for a 30-year period, one might expect the lender to be concerned that the collateral property might lose value due to sea-level rise related risks. We might expect that lenders would be wary of properties particularly susceptible to large-scale flooding due to storm surge associated with hurricanes, and either avoid making loans in such susceptible areas, or at the very least, charge higher interest rates and origination costs. But real estate agents have not observed such behavior on the part of lenders. This lack of response deserves some explanation.

As is the case with respect to the impact of flood risk on house price, it is also true that empirical studies of lender behavior in flood-prone coastal areas have reached conflicting results. In an analysis of more than 1.2 million conventional 30-year mortgages originated between 1992 and 2018, Nguyen et al. (2021) identified a

Fig. 8.2 Are Mortgage loans affected?

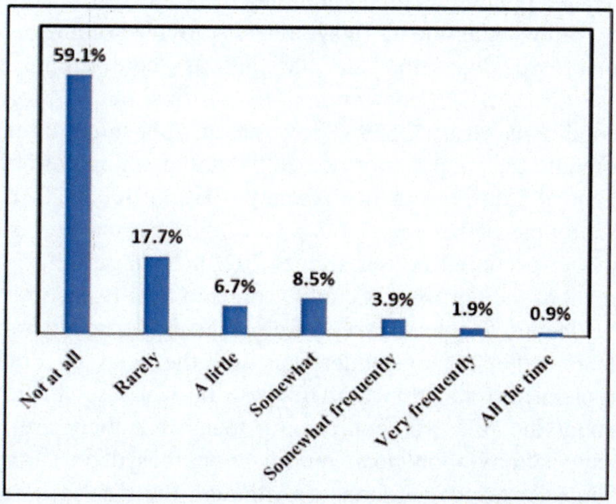

"sea-level rise premium" in interest rates for zip codes exposed to sea-level rise risk. Other studies, however, have suggested that lenders are not responding to risks of climate change or coastal flooding. In their assessment of the impact of sea-level rise on the housing market, Keys and Mulder (2020) concluded that although homebuyers are increasingly aware of the risks of coastal locations, the lenders have not been responsive to this risk: "our interpretation, informed by conversations with multiple lenders and industry participants, is that lenders are protected by federal programs that actively mis-price climate risk." Similarly, Ouazad and Kahn (2020) reasoned that it is not the lenders who are at risk, but instead it is the "government-sponsored enterprises" and specifically Fannie Mae (the Federal National Mortgage Association) and Freddie Mac (the Federal Home Loan Mortgage Corporation) that are assuming almost all the risk of mortgage default. They argue that the arrangement of lenders selling loans to Fannie Mae and Freddie Mac "incentivizes borrowers to locate to flood-prone areas" and "weakens the discipline brought about by the mortgage finance industry in fostering climate change adaptation." (Ouazad and Kahn 2020). We have already described the subsidization of climate risk that is attributable to current practices in the pricing of flood insurance through the National Flood Insurance Program. Does a similar shifting of risk exist for mortgage loans?

The answer to this question is "yes." First, it is helpful to share a bit of background about loan origination. Those who watch classic films may be reminded of the relationship between the local community and the mortgage lender portrayed in the 1946 film "It's a Wonderful Life." In that film the Bailey Brothers' Building and Loan company provided home loans to facilitate homeownership in the fictional town of Bedford. Any similarity between that image and the current mode of home mortgage lending should be set aside, however. The current largest home mortgage lender in the United States is an online lending site, Quicken Loans or Rocket Mortgage (Passy 2020). Rather than the cozy small-town relationship between lender and borrower portrayed in the film, loan officers associated with such online lenders operate remotely and may have little or no contact with borrowers.

But whether the mortgage loan is provided online or in-person, most lenders seek to provide loans that are classified as "conforming" so that it can be resold or "securitized." A "conforming" loan is one that "conforms" to the provisions or the conditions set by GSEs ("government sponsored entities") such as Fannie Mae or Freddie Mac and do not exceed the dollar amount established by the limit set by the Federal Housing Finance Agency (FHFA). As of 2021, the maximum loan amount set by the FHFA for all but a few high-cost markets is $548,250 for single family, owner-occupied houses (Segal 2020). Underwriting guidelines include an assessment of the borrower's credit score and debt-to-income ratio, as well as the loan-to-value ratio for the assessed value of the property. If a home loan is "conforming," the loan originator or lender can sell the loan to Freddie Mac or Fannie Mae, replenishing its supply of funds to be lent elsewhere. The FHFA is very specific about the rules for conforming loans, but there are no explicit requirements or expectations about loan modifications based on susceptibility to flooding, storm surge, or hurricane damage. Although the loan may require flood insurance if the property is in a FEMA-defined flood zone, the lenders do not adjust other loan conditions such as interest rates or origination costs for property in flood-prone areas.

Mortgage lenders intending to re-sell the mortgages to Fannie Mae or Freddie Mac therefore have no specific incentive to add new requirements concerning the borrower or collateral property on which the loan is based. But what about "non-conforming" mortgage loans? These loans make up only a small fraction of mortgage loans: in Florida, only about 6% of the first lien, owner-occupied 1–4 family home loans reported by HMDA for 2017 were non-conforming (Consumer Finance Protection Bureau 2021). These loans were ineligible for resale to a government sponsored enterprise but could be resold to other investors. Non-conforming loans include those made to borrowers whose credit histories are not quite up to the required level, or, more commonly, in amounts that exceed the FHFA maximum. Such loans are known as "jumbo mortgages." Lenders may tend to scrutinize these loans more carefully, requiring higher income levels or charging higher interest rates and larger origination fees (Nguyen et al. 2021).

But empirical research has demonstrated that even these loans are likely to be repaid in the event of a major disaster. In a study of Fannie Mae mortgages in areas affected by Hurricane Harvey, Kousky et al. (2020) found that "borrowers with flood insurance are more likely to prepay their mortgage, either because they can use insurance proceeds to pay off their mortgage, or because they are more willing to sell their damaged homes to investors since their losses are already compensated by insurance." Similarly, Gallagher and Hartley (2017) found that after Hurricane Katrina, homeowners were likely to use flood insurance to cover mortgage debt, sometimes under pressure from lenders.

In sum, there is little incentive for home mortgage lenders to take flood risk into account explicitly when making a home mortgage loan. If the loan is "conforming," it will be resold to Freddie Mac or Fannie Mae, relieving the lender of any risk associated with default. If it is not conforming, it still may be resold, but for loans to property where federal insurance is required, the borrowers are likely to repay the loan balance with their insurance pay-outs. It is understandable that mortgage lenders are not integrating climate change risks into their lending decisions since they have no long-term financial interest in the likelihood that the collateral property will suffer flood-related damage.

8.3 Do Appraisers Consider Elevation or Susceptibility to Flooding in Their Assessment of Value?

Appraisers, however, are in a different situation from lenders, and we might have expected more of a response from them. A home buyer, seller or real estate agent may seek an appraisal, but the appraised value of the property is most commonly obtained by the mortgage lender. The purpose of the appraisal is to provide an independent estimate of the home value so that the lender can decide on the amount and the terms of a mortgage loan. For residential real estate, the most common appraisal process involves a version of the hedonic price analysis considered in detail in Chap. 3. This analysis involves identifying a set of "comparable" houses that have sold recently and adjusting the estimated value of the house in question based on the relative age, size, number of bedrooms and bathrooms, the general condition of the house, and upgrades or home improvements, as well as other physical features such as views (Roulac 2007; Kalfrin 2020). Although these estimates are presumed to be bias-free, numerous studies have considered the accuracy of such appraisals and the ways informal matters, such as the relationship between the appraiser and the real estate broker or lender, can influence the final outcome (Eriksen et al. 2019; Hansz 2003; Agarwal et al. 2017; Ding and Nakamura 2016). What this means practically is that appraisers may align their estimates with the price on which the seller and buyer have agreed so that they can facilitate the sale.

Appraisers making recommendations for certain types of loans, such as those from the Veteran's Administration or the Federal Housing Authority have specific standards, for example checking for an absence of wood rot, termite damage or roof damage. However, the most important general responsibility of the residential real estate appraiser is to "investigate the characteristics of the subject property that might impact the property's value" (Appraisal Institute 2013). Does elevation matter? The Uniform Standards of Professional Appraisal Practice (Appraisal Foundation 2020) do not mention elevation as a criterion for assessing value. Obviously, flood hazard is part of the FEMA classification resulting in whether flood insurance is mandated, but this issue is not part of the specific process that assigns a value to a particular property.

Since real estate agents maintain a close relationship with appraisers, they are well-positioned to assess the response of appraisers to flood hazard. The question we posed was: "*Appraisers might also consider elevation or flood likelihood in their valuation of property. To what extent do you believe that property appraisers consider elevation or the likelihood of flooding in their assessment of home value*?" Real estate agents responded that appraisers were more likely than mortgage lenders to consider elevation or flood risk in estimates of market value, but nonetheless very few responded that this happens frequently, and the modal response was "not at all" (Fig. 8.3).

To summarize what we have found so far, real estate agents tell us few buyers avoid properties that will be susceptible to flooding as a product of sea-level rise, and that they have seen virtually no impact on sales price. Recent home sales at the high end of the market certainly suggest that there is continued demand for coastal property regardless of the risk of flooding due to sea-level rise. We mapped the locations of houses for sale in March of 2021 with list prices of $5 million or more (Fig. 8.4). Almost all of them are waterfront properties either directly on the coast or on a barrier island. These homes are almost exclusively located in areas within the 500-year floodplain, in areas especially susceptible to sea-level rise.

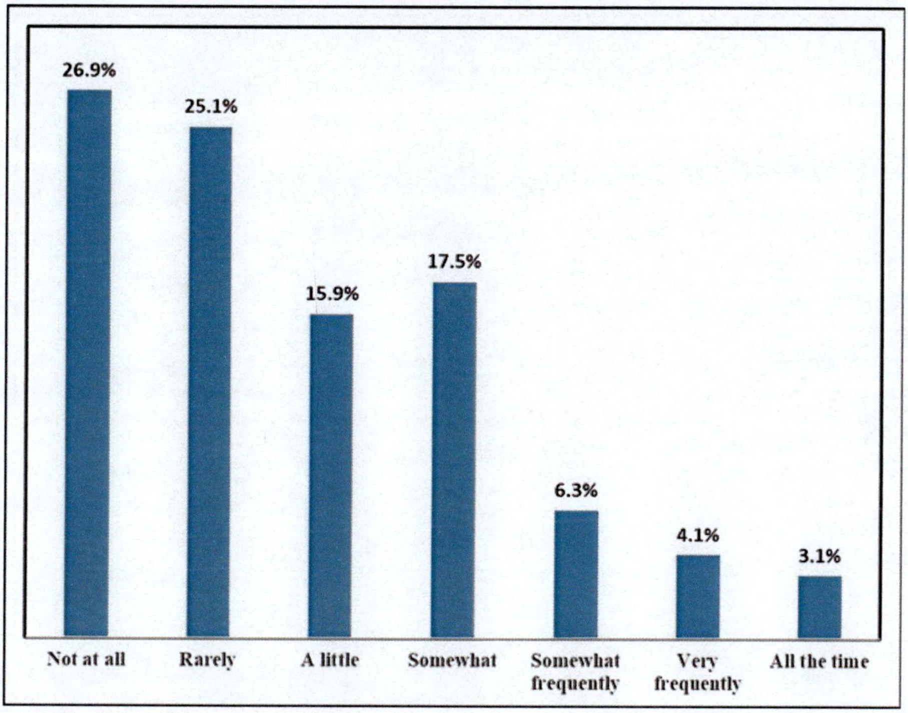

Fig. 8.3 Does elevation impact appraised value?

8.4 What Do Real Estate Agents Predict for the Next 5–10 Years?

We next wanted to explore how real estate agents see the future market for low-lying residential property. Do the real estate agents predict future lower demand and lower prices in areas prone to sea-level rise? We expected there would be variability in the predictions of real estate agents, partly based on differences in personal beliefs about sea-level rise and partly because of the experience the real estate agents have in the submarkets within which they were working. We also expected the real estate agents might be unduly optimistic about the future of flood-prone areas because such optimism serves their interests in selling houses. We asked: "*how do you see the near-term future for residential property in low elevation coastal areas? Will the residential property market see less demand and lower prices in areas prone to flooding in the next 5–10 years*?" Response options ranged from (1) no effect, (2) almost no effect, (3) very small effect, (4) small effect, (5) moderate effect, (6) large effect, to (7) very large effect of flooding.

Just as the real estate agents contended that there is little impact of flood risk on current house prices, they also said that they expect the same conditions will hold over the next 5–10 years (Fig. 8.5). Only 10% of respondents expect there to be a

Fig. 8.4 Properties listed for sale in April 2021 at prices of $5 million or more

large or very large effect of flood risk on demand or price for coastal housing, while almost 70% believe there will be little impact. Real estate agents' responses to this question varied somewhat by the submarket in which they were selling real estate (Fig. 8.6). Respondents from Miami-Dade County were more likely than the others

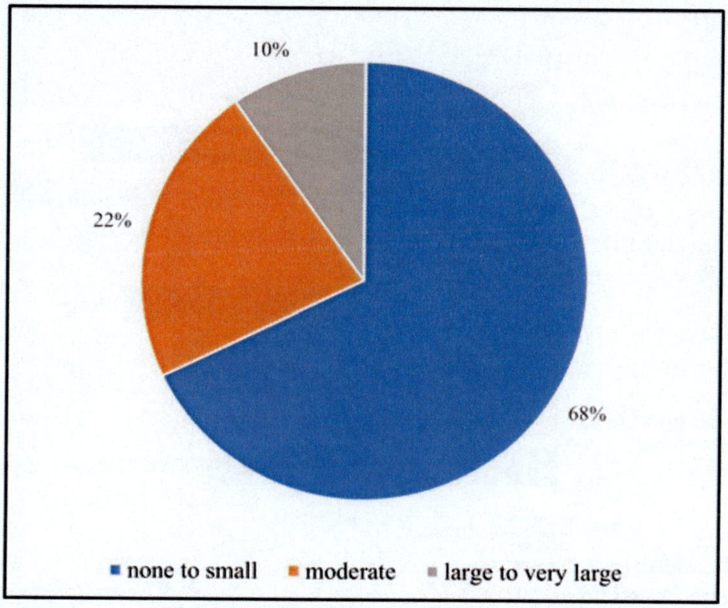

Fig. 8.5 Impact of flooding on property market over next 5–10 years for full sample

to predict that residential property in areas prone to flooding will see less demand and lower prices in the next 5–10 years, with more than a fifth expecting large to very large effects of flooding on the market.

8.5 In Their Words

Finally, many surveys including ours have tried to capture the complexity of opinions about housing markets with simple scales. We wanted to supplement these data with accounts from the real estate agents themselves. Of the 680 total respondents, 94 provided elaborations of their views of the impacts of sea-level rise on demand and house prices in the next 5–10 years. To elicit these responses we posed these questions: *"Do you think that sea-level rise will weaken demand for coastal property in your market area? Why?"*, and *"Do you think that sea-level rise will result in lower prices for coastal property in your market area? Why?"*

About 70% of these respondents indicated that there would be little-to-no impact of sea-level rise on either demand or price in the next decade. Most of them based this opinion on their current experience with high demand for coastal property. They described the current buyer motivations this way:

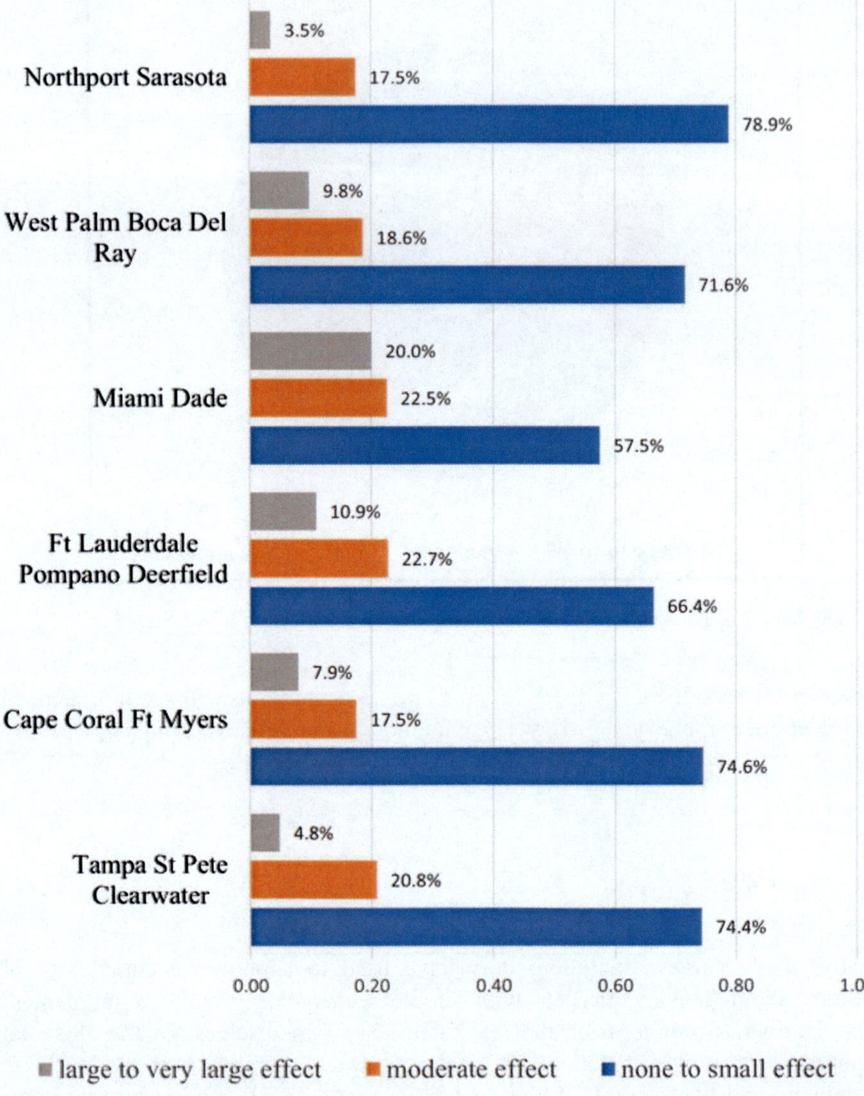

Fig. 8.6 Future effect of flooding by submarket

- *There is always a limited supply of water property with more buyers than sellers. At the end of the day people will always live the lifestyle they want to live.*
- *Prices on the coast are holding strong, there is so much talk on living the Florida lifestyle and very little on sea-level rising.*
- *Florida waterfront real estate continues to draw legions of buyers, prices continue to rise and the availability of waterfront property continues to diminish. If*

sea-level rise truly exceeds, strangely enough, the Russian climatological estimates, which happen to be the only estimates that can be called reliable and fairly accurate, then we are over a century away from widespread sea-level rise along the Florida coastline. In that time, science, technology, engineering methodologies and practices will surely address and mitigate it. Florida waterfront real estate is a treasure that will always be protected. It's simply too beautiful, too scarce and too valuable.

- *I do not think that sea-level rise will weaken the demand for coastal properties in our market here. People that purchase waterfront property are water people. They like to fish, boat, or just relax or meditate on the water. Not being a water person myself, they may justify this by saying it's better than the tornadoes in the Midwest, earthquakes & fires in California, etc. Anything can happen anytime & anywhere!*
- *People are and will still buy in the coastal areas of Florida and if they are buying there will be no decrease in value. This market driven dynamic is unique due to several factors. The largest pool of buyers driving market are retired or soon to be retired people and they have the belief that they will be long gone before there is any impact from climate change. They mainly are buying on emotion and not factoring in the long-term cost of ownership. They are also buying with cash and no mortgage.*
- *It's Florida and buyers want to be near/on water. Buyers of coastal properties are financially able to be more risk-oblivious and can afford the higher rate for insurance or be self-insured. Sea-level rise is not currently top-of-mind in our local market.*
- *The market is red hot and with COVID conditions (and northern states being shut down) people are flocking to southwest Florida, purchasing property or even simply renting and working from their homes. Cash buyers are not concerned about flood insurance. Strong state leadership is allowing Florida businesses to flourish and the demand for homes and jobs is not going to slow due to sea-level rise in the next 5–10 years. (It may slow due to other factors we are not aware of yet).*

A second large group of respondents agreed that neither demand nor prices would be affected, but qualified their statements by mentioning adaptations or mitigations that would protect new housing from rising waters:

- *I do not think that sea-level rise will negatively affect demand for waterfront properties. Many buildings in Fort Lauderdale and Miami have their lower floors flooded during hurricanes. The response to that is to raise the lobby to the second or 3rd floor and/or build stairs or stilts on the ground floors. This has not weakened demand for coastal property. Buildings are being razed and rebuilt to adapt to the flooding.*
- *Engineers are ingenious, will figure out how to build levees, but sea-level rise is overblown anyways.*
- *The type of buyer that can afford coastal properties may signal publicly that they are concerned but privately do not care about land erosion. For the few homes*

that have had this issue, the seller will make fortifications (strengthen the seawall or have sand brought in) to maintain the esthetics and property value. Homes on the West Coast are literally falling into the ocean or burning down, and buyers are lining up to feast on the spoils.

- *I think that in ten years when people see and feel that the effects of sea-level rise will be more palpable, and then the demand will weaken. However wealthy people will still be enamored by the idea of living in front of the sea, but they will probably spend a lot of money making the property more resilient to the effects of sea-level rise. This means that maybe the demand for high end properties will not weaken so much.*
- *Owners will just build with higher floor levels to protect property. This is what has happened in the past. Recent higher floor levels have been due to flood insurance demands & new construction.*
- *If the sea-level rises, technology will be there too to adjust the way that houses near water will be built.*

A third group felt that there might be some response in terms of demand and price at some point, but that we will not see this effect any time soon:

- *I don't think the sea-level rise will be high enough to make a noticeable difference in our lifetime. And in longer, geological timeframes, the Earth has gone through many cycles of Ice Age and Greenhouse Ages.*

  People only respond to pain (when it's happening). Unless it is a specific property and one cannot obtain flood insurance, I don't see it affecting prices.
- *People are not informed. All people care about is the dream which is waterfront!*
- *People rarely think to the future. Further, out-of-state buyers rarely are familiar with the hazards of local properties.*
- *At some point, weakening demand and increased inventory from sellers not wanting to be caught with the hot potato should lead to lower prices. But this is likely not years but decades down the road. Public perception and understanding of sea-level rise is simply not there yet.*

Additionally, several of the respondents who predicted no impact on demand or price indicated that they do not accept the notion of climate change or sea-level rise:

- *It's all an unproven scientific story by Al Gore to make more money*
- *There is no appreciable RISE (FALSE FACTS) and Global Warming is not happening. Global Change (currently colder) is natural and the levels of change are a MYTH.*
- *We are in fact heading into a minor ice age regardless of all the politically motivated climate alarmists.*

Only ten respondents replied that both demand and house prices will be affected at least to some extent by sea-level rise. Of these, three attributed the impact of sea-level rise to an increasing cost of flood insurance in high-risk areas:

- *I believe that the insurance premiums will affect the demand. At the same time some neighborhoods will suffer a decline in demand, but it will be isolated cases in my opinion.*
- *I think that the sea-level rise will dampen demand for waterfront properties because insurance rates will rise, and when demand drops, prices drop.*
- *We may see flood insurance regulations change to make it very expensive to obtain flood insurance coverage. Should this coverage become much more expensive, it will have a negative effect on the value of waterfront homes. I don't think sea-level rise, in and of itself, will be the cause for any decrease in demand, but the ancillary effect of rising costs of insurance may have that effect.*

The others acknowledged a direct effect of sea-level rise on decreasing attractiveness of vulnerable property to homebuyers and therefore a shift in demand and a reduction in price. The reasoning was sometimes complex, and therefore has been replicated without redaction.

- *Yes, more people are moving inland and away from the flooding. Yet close enough to enjoy downtown Miami. More development in areas with higher ground like Liberty City, Little Havana, and Little Haiti. With beachfront properties costing much more, there is more demand going to new build areas further away from the coast.*
- *In the longer term, unless humanity alters our collective impact on the environment, it is quite likely that an "event" may trigger a weakening of demand for coastal property and hence prices. We have a 6-month hurricane season and in November, we have the "King tides" where the sea-level rises dramatically. Either one of those could create a flooding of the barrier islands and/or low-level waterfront property. This would result in a dramatic hike in flood insurance which in turn would make the waterfront homes a more costly asset to maintain. However, it is hard to be certain that this would lead to "lower prices" in absolute terms, as money supply has been so loose over the past decade, we may see inflation creeping back into the equation. Nevertheless, on a "relative" basis, it would seem quite likely that coastal properties may underperform other areas that are not at such high risk. There is already evidence that "higher ground" neighborhoods such as Little Haiti in Miami, which sits on a coral ridge, have outperformed other local downtown areas, in what is being called "environmental gentrification." Investors, like myself, have tried to position themselves for what seems to be an inevitable continued rise in sea-level for the foreseeable future. As a zero-income tax State, Florida is dependent upon annual real estate taxes whereby each home is "appraised" annually. Fort Lauderdale is sometimes called "the Venice of the Americas" so the majority of the annual tax base would be from higher-priced waterfront homes. Therefore, the State's economic viability is largely tied to maintain high appraisal values.*
- *I believe that the sea-level rise could result in weakened demand for homes directly on the ocean at or near sea-level. Homes built on higher elevations*

and further away from the high-water mark would still be desirable to many buyers.

- *There are still a large percentage of people that are willfully ignorant of how the sea-level rise will impact properties in Florida. I suspect much of this is political as people believe what they hear in their echo chamber. Once politicians are forced to discuss multi-billion-dollar projects to try and "hold back the sea," property owners and buyers will be shocked to see the subsequent rise in property insurance. This will only accelerate during the next 50–100 years.*

The responses of these real estate agents reveal their commitment to Florida as a destination and their appreciation of the Florida coast. They refer repeatedly to the beauty of the coast, as well as the lifestyle that accompanies this location. As one of them put it: "*demand for coastal property is based upon quality of life. Boating, sun, surf sand and a resort style living. This is something that we in the US market as the pinnacle of lifestyle.*" In general, real estate agents are optimistic about the future of coastal Florida, and undoubtedly convey this enthusiasm to homebuyers.

8.6 Impact of Demographic, Locational and Attitudinal Characteristics on Responses

In Chap. 7, we reviewed the impact of the demographic, locational and attitudinal characteristics of real estate agents on their interpretation of whether clients are avoiding coastal properties. The same characteristics could also impact their opinions about whether house prices are falling or stagnating in low-lying properties, whether lenders are more reluctant to make mortgage loan commitments in these areas, and if appraisers consider the negative consequences of location in a flood zone when setting a price value to the house. They could also affect their views of the future of the real estate market in Florida. We therefore calculated ordinary-least squares (OLS) models to assess factors associated with variance in real estate agents' responses.

We used the same model of demographic, locational and attitudinal characteristics that we included in the OLS regression models in Chap. 7, with a cut-off of <0.05. We found **no** variables predictive of variation in real estate agents' responses about whether prices are falling (Table 8.1). In other words, regardless of their personal characteristics, beliefs or political orientation, real estate agents hold that prices are neither falling nor stagnating.

With respect to real estate agent experience with lenders and appraisers as well as their expectations about the 5–10-year future of the housing market, the demographic, locational and attitudinal correlates were similar to those for beliefs about buyers avoiding flood-prone areas as presented in Chap. 7. Real estate agents who were older, who had sold fewer houses in the past year, who agreed that climate change and sea-level rise are taking place, and who were located on the Atlantic coast were more likely to report that lenders had made modifications in loans to

Table 8.1 Real estate agents' characteristics and beliefs about house prices falling, lender response and appraiser response to Property at risk of flooding

	(1)	(2)	(3)	(4)
	Price fall	Lenders	Appraisers	Fut. demand
	Coefficient	*Coefficient*	*Coefficient*	*Coefficient*
Age	0.03	0.16**	0.18**	0.16**
	(0.05)	(0.05)	(0.06)	(0.06)
Education	−0.02	0.01	0.02	−0.02
	(0.04)	(0.05)	(0.05)	(0.05)
Female	0.01	−0.07	0.26	0.09
	(0.10)	(0.11)	(0.13)	(0.13)
Minority	0.26	0.17	0.33	0.22
	(0.14)	(0.15)	(0.17)	(0.16)
Republican	0.17	0.01	0.34*	−0.14
	(0.12)	(0.14)	(0.16)	(0.14)
Democrat	0.07	−0.12	−0.09	−0.01
	(0.14)	(0.15)	(0.17)	(0.16)
Sea-level scale	0.05	0.09*	0.12*	0.23***
	(0.04)	(0.04)	(0.05)	(0.05)
Years in Florida	0.09	−0.04	−0.09	0.07
	(0.05)	(0.05)	(0.07)	(0.06)
Houses Sold	−0.20	−0.34**	−0.46**	−0.28*
	(0.11)	(0.13)	(0.15)	(0.14)
Atlantic	0.18	0.24*	0.27	0.23
	(0.11)	(0.12)	(0.14)	(0.13)
Constant	1.09*	0.74	1.33*	1.25*
	(0.48)	(0.48)	(0.61)	(0.53)
N	*612*	*608*	*609*	*610*
Adjusted R^2	0.015	0.039	0.050	0.080
AIC	1997.2	2080.9	2287.5	2211.7
BIC	2045.8	2129.4	2336.1	2260.3

Note: Cell entries are OLS coefficients with standard errors in parentheses below.
$^{*}p < 0.05$, $^{**}p < 0.01$, $^{***}p < 0.001$

prospective buyers in flood-prone, low-elevation areas (Table 8.1). Respondents who reported that appraisers took elevation into account in the valuation of property tended to be older, had sold fewer houses, were Republican, and believed in climate change and sea-level rise. Finally, real estate agents who were older, sold fewer houses, and agreed that climate change and sea-level rise are taking place were also more likely to believe that flooding will affect the housing market in the next 5–10 years.

Other factors may also be at work in explaining real estate agents' views of current and future house prices. First, it is in the best interest of the real estate agent for the housing market to remain strong with large numbers of sales and ever-increasing prices. The phenomenon of motivated reasoning reviewed in Chap. 4, that is, the interpretation of observations based on what is preferred rather than what is occurring, may affect the judgements of the real estate agents. In addition, one of the desired characteristics of any salesperson is optimism, whether or not this optimism is justified by facts. It therefore could be the case that although property values are actually being affected by the elevation of the property, the real estate agent has reason to not acknowledge such trends and continue to hold the view that, in this case, climate change and sea-level rise currently has and will continue to have no impact on the Florida housing market.

A second factor affecting real estate agent response to the market is their awareness of changes in the building codes to increase the safety of new property from flood hazards. As some of the respondents noted, building code modifications such as the requirement to elevate the first floor will mitigate some of the risk for homebuyers who want a coastal location and its associated lifestyle. Florida Building code as modified in 2020 requires new residences in "flood hazard areas . . . shall have the lowest floors elevated to or above the base flood level plus 1 foot." (Florida Building Code 2020). The base flood level has a technical definition but refers to an area with at least a 1% chance of flooding each year as specified by the Flood Insurance Rate Map (FIRM). The non-elevated portion of the house may be used for parking and storage, but the living space must be elevated to meet state and local building requirements. New home construction is built to comply with this code, and some older homes have been modified to raise their lowest floors to conform to this building code (Fig. 8.7). Miami Beach, Tampa and St. Petersburg have also established local funding to help homeowners elevate existing structures, and homeowners whose houses were damaged by recent hurricanes such as Hurricane Irma in 2017 can apply for a federally funded, state-managed fund (Rebuild Florida) to repair and mitigate against future storms, which includes elevating the structure.

Third, the real estate agents are likely well aware of efforts by some of the communities to protect infrastructure from sea-level rise. Miami Beach, for example, invested $600 million to raise roads and improve stormwater damage. They also made changes in their land use and development codes, introducing "green infrastructure" to absorb water with vegetative systems such as mangrove planting. These types of measures are likely not only to continue, but also to increase in frequency and comprehensiveness with increased awareness of the risks of sea-level rise. The existence of both land use mitigations and changing structural requirements may contribute to the optimism of real estate agents about the future prospects for coastal properties.

Fig. 8.7 Example of an elevated dwelling. (Source: FEMA, Coastal Construction Manual)

An additional explanation for the observations of the real estate agents may relate to the timing of market changes. Keys and Mulder (2020) suggested that there was a time lag between a decrease in demand and the impact of this decrease on prices. The real estate agent response to house prices may be reflecting the immediate past more than the immediate future with respect to price trends and were we to return to the same population in a few years and re-interview them, we might get different results.

Finally, the real estate agents may simply have a more nuanced understanding of the interaction between potential coastal flooding, sea-level rise and the housing market.

8.7 Summary

Our survey of real estate agents found no evidence that relative to property that is inland or at higher elevation, prices for flood-prone coastal property in Florida are stagnating or declining. Some respondents reported that there are some buyers who wish to avoid flood-prone properties, primarily because of the cost of mandatory flood insurance. However, they observe that more buyers still desire low-lying coastal property for the beach lifestyle it provides. Most of the real estate agents reported little or no negative impact of such location on selling price, and instead note that beachfront properties are in high demand.

When asked to speculate about property markets over the next 5–10 years, the respondents showed a bit more caution, and particularly those real estate agents who themselves are convinced that the climate is changing, and that the sea-level is rising.

But as we have noted previously, the housing market in South Florida is far from monolithic. There are sharp differences in types of buyers, planned housing tenure, and general home prices from one region to another. Even the real estate agents show systematic regional differences with respect to political orientation and beliefs about climate change, and the relationship between these beliefs and their perceptions of housing market trends. Further, historic development practices have left some areas far more vulnerable to flooding and sea-level rise than others. The next chapter will explore some of this variability and its implications for these areas.

References

Agarwal S, Ambrose B, Yao, V (2017) The effects and limits and regulation: appraisal bias in the Mortgage market. SSRN Working Paper

Appraisal Foundation (2020–2021) Uniform Standards of Professional Appraisal Practice (USPAP). Available at: https://www.appraisalfoundation.org/imis/TAF/Standards/Appraisal_Standards/Uniform_Standards_of_Professional_Appraisal_Practice/TAF/USPAP.aspx?hkey=a6420a67-dbfa-41b3-9878-fac35923d2af

Appraisal Institute (2013) Understanding the Appraisal. Chicago, Available at: https://www.appraisalinstitute.org/assets/1/7/understand_appraisal_1109_(1).pdf

Consumer Financial Protection Bureau (2021) Home mortgage disclosure act data. Available at: https://www.consumerfinance.gov/data-research/hmda/historic-data/

Ding L, Nakamura L (2016) The impact of the home valuation code of conduct on appraisal and mortgage outcomes. Real Estate Econ 44(3):658–690. https://doi.org/10.1111/1540-6229.12120

Eriksen MD, Fout HB, Palim M, Rosenblatt E (2019) The influence of contract prices and relationships on appraisal bias. J Urban Econ 111:132–143. https://doi.org/10.1016/j.jue.2019.04.007

Florida Building Code (2020) Alterations—level 1. In: Existing Building (7ed). Available at: https://codes.iccsafe.org/content/FLEBC2020P1/chapter-7-alterations-level-1

Gallagher J, Hartley D (2017) Household finance after a natural disaster: the case of Hurricane Katrina. Am Econ J Econ Pol 9(3):199–228. https://doi.org/10.1257/pol.20140273

Hansz JA (2003) The use of a pending mortgage reference point in valuation judgment. J Prop Invest Financ 22(3):259–268. https://doi.org/10.1108/14635780410538177

Kalfrin V (2020) From the outside in: what do home appraisers look for in a house? In: Homelight. Available at: https://www.homelight.com/blog/what-appraisers-look-for-in-a-house/

Keys BJ, Mulder P (2020) Neglected no more: housing markets, mortgage lending, and sea-level rise, No. w27930. National Bureau of Economic Research. https://doi.org/10.3386/w27930

Kousky C, Kunreuther H, LaCour-Little M, Watcher S (2020) Flood risk and the U.S. housing market. J Hous Res 29:S3–S24. https://doi.org/10.1080/10527001.2020.1836915

Nguyen DD, Ongena S, Qi S, Vathunyoo S (2021) Climate change risk and the cost of mortgage credit. Swiss Finance Institute Research Paper Series 20–97. https://doi.org/10.2139/ssrn.3738234

Ouazad A, Kahn ME (2020) Mortgage finance in the face of rising climate risk. National Bureau of Economic Research Working Paper No. 26322. Available at: https://www.nber.org/system/files/working_papers/w26322/w26322.pdf

Passy J (2020) Quicken loans is going public: 5 things to know about the mortgage lender. In: Marketwatch. Available at: https://www.marketwatch.com/story/quicken-loans-is-going-public-5-things-to-know-about-the-mortgage-lender-2020-07-08

Roulac SE (2007) Brand+ beauty+ utility= property value. Prop Manag 25(5):428–446. https://doi.org/10.1108/02637470710824711

Segal T (2020) Conforming loan limit. In: Investopedia. Available at: https://www.investopedia.com/terms/c/conformingloanlimit.asp

Weintraub E (2020) Using your agent's recommended mortgage lender. The Balance. Available at: https://www.thebalance.com/using-your-agent-s-recommended-mortgage-lender-1798449

Chapter 9
Four Communities: Vulnerable by Design

> *From its founding as an imperial outpost to its modern identity as a tourist empire, Florida has evoked contrasting and compelling images of the sacred and profane: a Fountain of Youth and Garden of Earthly Delights, a miasmic hellhole and scuzzy wasteland.*
>
> Mormino, G. R. (2005, p. 3)

Abstract Many coastal Florida real estate developments were created by draining wetlands, dredging the land to create canals, and using the dredged sediments to create homesites at elevations only slightly above sea-level. Some of the original developers had idealistic intentions of creating garden communities, but most were trying to maximize profits by creating as much waterfront property as possible. Current residents of these communities are now living in areas susceptible to a combination of subsidence and sea-level rise. This chapter reviews the origins of four such communities: Apollo Beach, Cape Coral and Punta Gorda, on the Gulf Coast, and Hollywood on the Atlantic Coast. Each is highly vulnerable to flooding primarily because of the way the land was created and platted and each is therefore "vulnerable by design."

Keywords Dredging · Property development · Planned communities

There is much that the various communities in South Florida share. They owe their settlement to the extension of the railway system to once-remote areas, as well as the development of a large middle-class who could afford a vacation away from home or even the purchase of a second home in a subtropical setting. These communities welcome large numbers of tourists from throughout the world attracted to, among other features, the beautiful beaches as well as world-class theme parks. New permanent residents come to these communities each week. Some come from Latin American countries such as Cuba, Venezuela, Nicaragua, Colombia, or Mexico attracted by economic opportunities along with the opportunity to "keep familiar cultural roots" (Fajardo 2016; Enterprise Florida 2017). Other new Floridians come from Caribbean origins such as Puerto Rico, the Dominican Republic, or Haiti, or from colder climes of the American northeast or Midwest or from Canada. New

© The Author(s), under exclusive license to Springer Nature Switzerland AG 2022

R. Palm, T. Bolsen, *Housing Market Response to Sea-Level Rise in Florida*, Coastal Research Library 37, https://doi.org/10.1007/978-3-030-88435-2_9

residents come seeking freedom from oppressive regimes elsewhere, economic opportunity, or a place to live and retire with a beach lifestyle. What they are NOT coming for is the disappearance of their new home from storm surge or flooding caused by sea-level rise.

To be sure, not all of Florida is at risk of being submerged in the next century, or as John Englander (2020) put it, all of "Florida is Not Going Underwater." Counties with the highest elevations include Walton and Washington counties in the "Northern Florida Highlands" with elevations up to 345 feet, and in Lake County along the Lake Wales ridge in central Florida with elevations up to 312 feet. Our study area, however, focused on a survey of real estate agents doing business along the low-elevation coast. In this part of Florida, several communities having no point at an elevation above 10 feet (Fig. 9.1). As we have noted, these areas remain attractive to homebuyers. A systematic study of vulnerability to sea-level rise in southeast Florida (Bloetscher et al. 2016) found that although "wealthier people live in more geographically vulnerable areas" at present, the future will see an "increase in vulnerable land area that affects a larger and more diverse population."

How did the very process of land development in South Florida leave its residents vulnerable to sea-level rise? To answer this question, we will examine the responses to our survey from real estate agents whose offices are located in four highly vulnerable communities: Apollo Beach, Punta Gorda and Cape Coral on the Gulf coast, and Hollywood on the Atlantic coast (Fig. 8.1). In each of these communities, the developer had a vision of creating a community that would be attractive to new residents, particularly those who wanted to take advantage of the Florida lifestyle in a subtropical environment, who wanted to own a boat and to be close to the water. These developers dredged the pre-existing wetlands to create canals and raise the elevation of the land so that they could build new houses. Each of them destroyed the pre-existing wetland or mangrove environment that had provided some protection against flooding. And, with current rates of global warming, much of this waterfront property will experience severe flooding before the expiration of any 30-year mortgage loan.

9.1 Apollo Beach

Apollo Beach was first developed by Paul Dickman who intended to build a waterfront community for 40,000 people located only 12 miles south of Tampa (Maschek 2008; Nealen n.d.; Osprey Observer 2018). Dickman hired a Tampa engineer to begin dredging the mangrove waterfront, but sold the 5500 acre parcel in 1957 to a Michigan developer Francis J. Corr. Corr, and eventually his sons, began building and selling houses. Much like the development of Fort Lauderdale, the dredged land was formed into "finger wedges" with so that a maximum number of properties could have waterfront locations (Fig. 9.2).

An unincorporated place within Hillsborough county, Apollo Beach currently has a population of almost 21,000, having grown by 50% in the past decade (World

Fig. 9.1 Properties at a 1% risk of flooding in 30 years. Areas shaded in purple represent 3 or more feet of flooding. (Source: Flood risk data is provided by Flood Factor®, a product of First Street Foundation®)

Population Review 2021). Its population is 81.5% white with a median age of 45 and a homeownership rate of 85%. According to the Redfin website, the property market is "very competitive," which they define as many homes for sale getting multiple

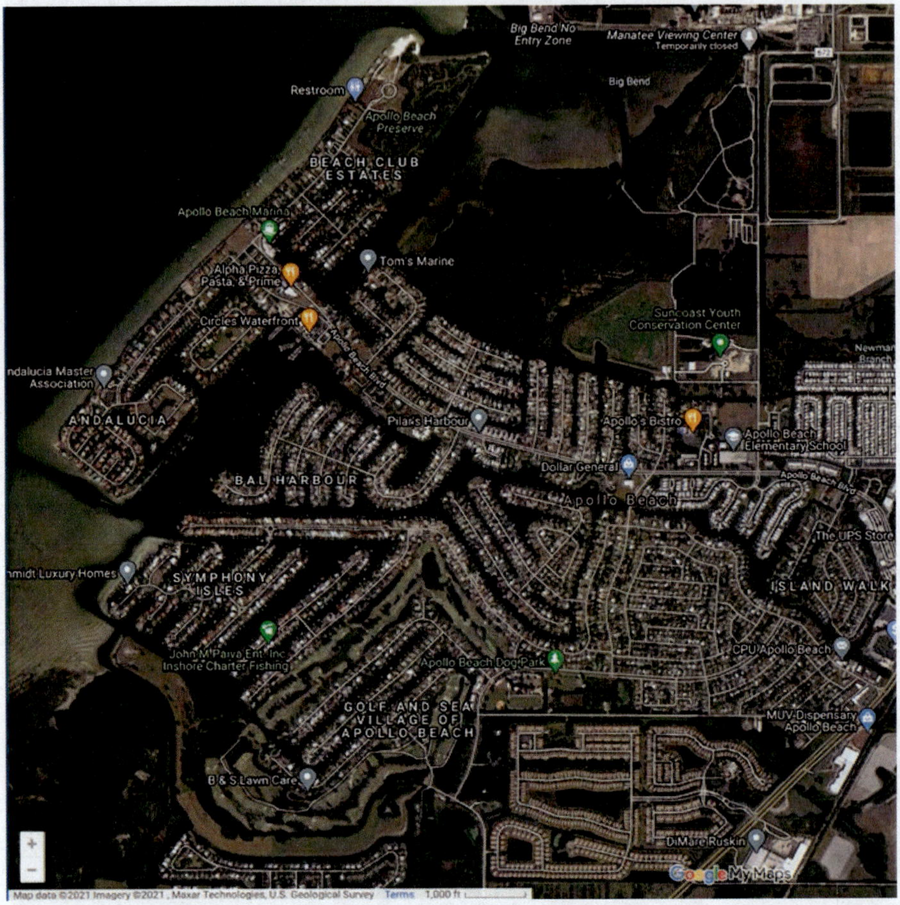

Fig. 9.2 Apollo Beach

offers and the average sale taking only 13 days (Redfin 2021). Prices of houses for sale in February 2021 ranged in price from $185,000 to $2.3 million (Table 9.1).

Since the elevation of Apollo Beach is 3 feet above sea-level, most properties are highly susceptible to sea-level rise induced flooding. When asked whether homebuyers are avoiding low-elevation properties, one broker answered "Some are avoiding properties in flood zones. Others accept the fact that if you live on the water, you're living at a low elevation." Three others mentioned the cost of flood insurance and noted that prospective buyers do not avoid flood-prone areas "as long as flood insurance is cost effective." One respondent added "typically buyers want to enjoy proximity to the water, so they care less."

Table 9.1 Apollo Beach

Apollo Beach	
Population (2021)	21,000
Median age	45 years
Race	81.5% White
Average home price	\$358,500
Range of home prices	\$185,000 – \$2.3 million
Average elevation	3 feet

9.2 Cape Coral

Cape Coral in Lee County is frequently described as a "planned community." This term is usually reserved for areas developed under a master plan with the goal of creating both a physical and social space for "ideas of community and the good life" (Gwyther 2006; Bartling 2008). At least at its inception, it is not clear that Cape Coral would fit the classic definition that such a term implies. As described in Chap. 2, it was developed and marketed in the 1950s by two brothers from Baltimore, using national advertising and high-pressure sales techniques to sell property on the installment plan (Mormino 2005). The developers platted thousands of lots for single family houses, dredged the wetlands and created 400 miles of canals which they could advertise as exceeding the canal mileage of Venice, a "waterfront wonderland." They used the sediments to create building lots that had an elevation of 5.5 feet (Williams 2017). The developers invited prospective buyers to a free dinner or even a charter flight to the property where they were subjected to high-pressure sales tactics. They also advertised by giving away plots of land on television programs such as *The Price is Right* and *Queen for a Day*. What the developers did not provide, at least in the early years, was the infrastructure around which a community could grow: "they also neglected to build Cape Coral any water or sewer infrastructure, and it has no direct access to the interstate—or even a good swimming beach—so it's struggled to attract businesses" (Grunwald 2017).

Presently, Cape Coral's population is 204,549, having grown by 33% since 2010 (World Population Review 2021). The median age of the population is 46.7, with about 1/3 of the residents over age 60. About one-fourth of its residents were born in Florida, and 65% were born elsewhere in the United States. It is 90% white, and about 75% of the housing units are owner-occupied with a median value of just under \$230,000 (World Population Review 2021).

At an average elevation of 5.5 feet, Cape Coral is highly vulnerable to storm surge. However, new houses are being built and sold as the city continues to grow.

Table 9.2 Cape Coral

Cape Coral	
Population (2021)	204,549
Median age	46.7 years
Race	90% White
Average home price	$230,000
Range of home prices	$105,000 – $3.9 million
Average elevation	5.5 feet

As of February 11, 2021, realtor.com listed 1726 single-family houses for sale in Cape Coral, ranging in price from $105,000 to $3.9 million (Table 9.2). As in other coastal Florida cities, the most expensive houses are almost always near the coast, in this case in the southern part of the city near the Caloosahatchee River (Fig. 9.3). This portion of the city is highly vulnerable to flooding associated with sea-level rise. Cape Coral real estate agents reported that buyers are "somewhat" likely to avoid low elevation property, but, like the others, found no evidence of prices being negatively affected by vulnerability to flooding. They also predict only a small effect of flooding on demand or prices over the next 10 years: "*we are in Florida and know that flooding more than likely occurs*" and "*flooding will be the next generations' problem*." They noted that "*lots of buyers here want waterfront homes*" and "*no one is worried about flooding because the water can only get so high and then it runs into the gulf.*"

9.3 Hollywood

Although Cape Coral can be understood as an attempt by developers to amass a fortune by selling off property in what was a largely unplanned community, Hollywood was a "planned paradise" established by Joseph Wesley Young, Jr. (Mickelson 2015). Its city plan including a canal, two artificial lakes, parks, schools, churches, and a golf course (Virginia TenEick 1989). Young invested millions of dollars to build what he called a "Dream City in Florida" with beautiful ocean beaches and artificial lakes (Fig. 9.4). His inspiration was the then-popular "city beautiful movement" of the late nineteenth and early twentieth century that proposed city design, the creation of green spaces and the construction of plazas and wide avenues could enrich lives and encourage a more civil society (Smith 2009; Fairfield 2018). In 1925 the city was incorporated, and Young was elected mayor.

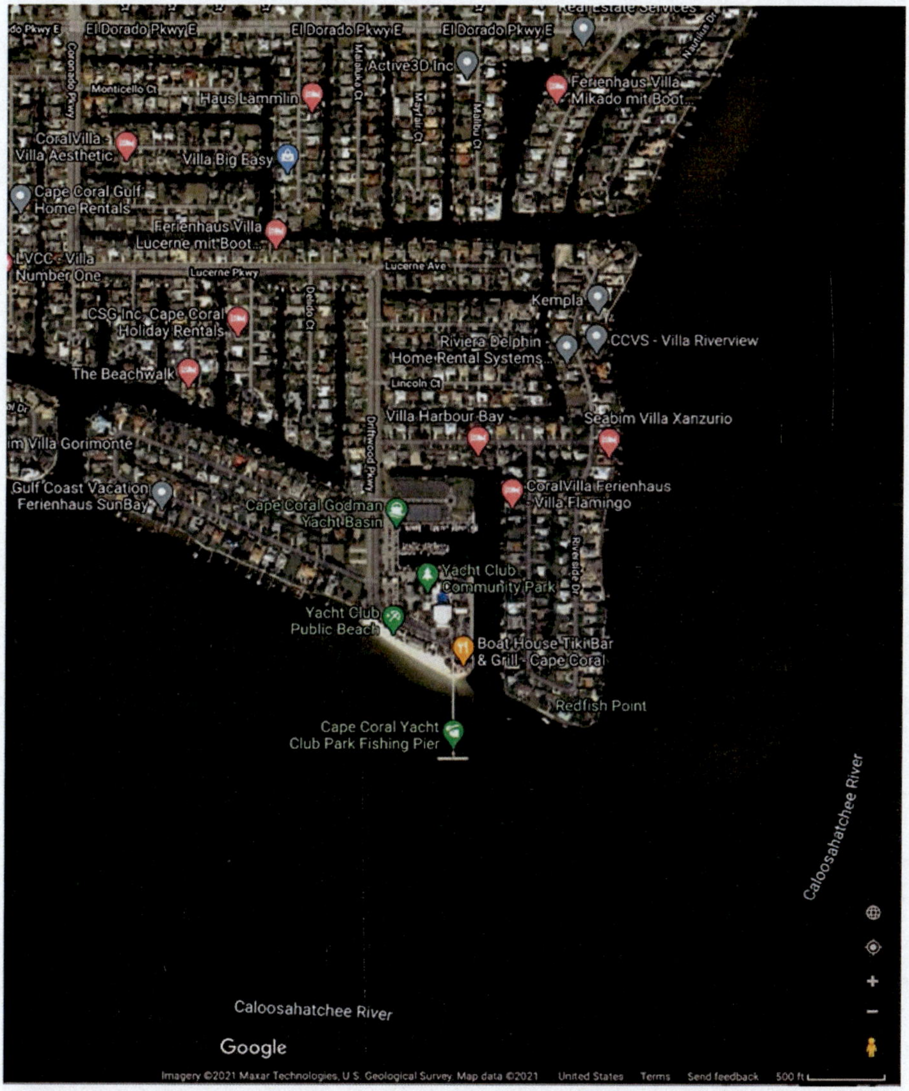

Fig. 9.3 Cape Coral

At first, population growth was rapid as Hollywood was marketed to New York City buyers. But the trajectory of growth was hampered by a devastating hurricane in 1926 as well as the stock market crash of 1929. Growth resumed in the 1960s when the population swelled from 35,237 to 106,873 (U.S. Census 2019).

Hollywood has been growing slowly since 2000, with a current estimated population of 155,689. Of its current population, 38% are foreign-born, primarily from Latin America. The World Population Review notes that it has one of the "highest populations of Cubans, Peruvians and Romanians in the country." About 68% of the

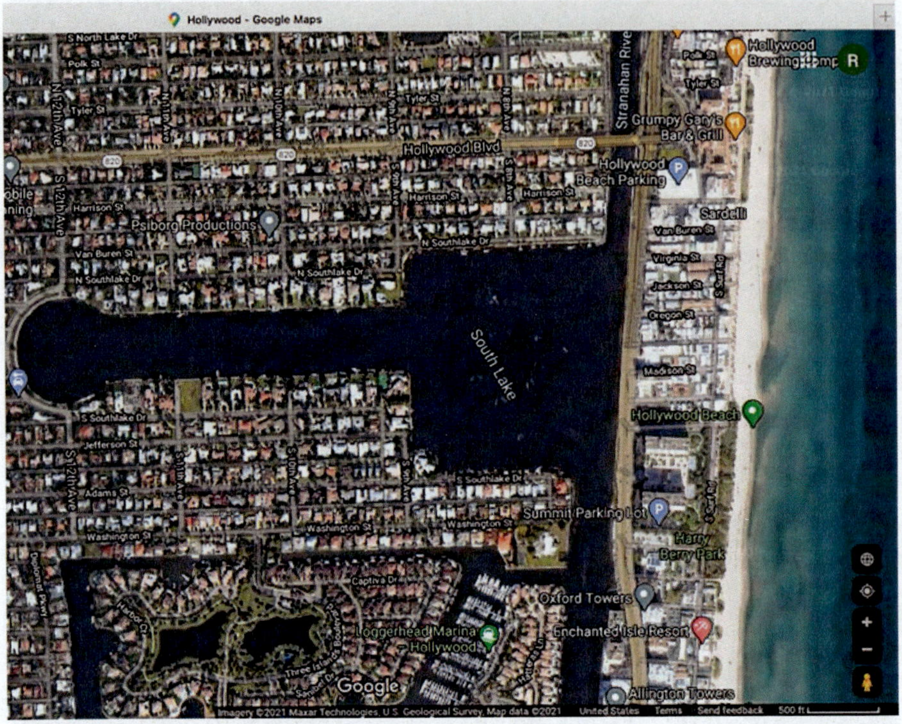

Fig. 9.4 Hollywood, Florida

population is white and 18% Black or African American. About 56.6% of the population own their own homes.

Redfin assesses the current housing market in Hollywood as "not very competitive", meaning that multiple offers are rare, houses sell at about 4% below the list price and are on the market for 66 days (Redfin 2021). The average sales price is $347,000. As of February 11, 2021, realtor.com listed 563 single-family houses for sale, ranging in price from $44,000 to $7.5 million (Table 9.3). The average elevation of Hollywood is 9.8 feet, and the elevation of Hollywood Beach is about 3 feet. The highest price houses are concentrated on the Hollywood Beach barrier island, an area particularly susceptible to sea-level rise-related flooding.

Hollywood real estate agents were more likely than those in Apollo Beach or Cape Coral to believe that the real estate market of the 5-to-10-year future would be affected by flooding, but also reported that at present few are avoiding such property or that prices were negatively affected. When asked whether buyers were avoiding low-elevation property, they gave responses such as "*no one thinks that far out,*" but also that buyers want to "*save money because of the high cost of flood insurance.*" One also noted that information provided by the Flood Factor maps is "*helpful to know as a buyer or seller but has little weight in the decision-making process.*"

Table 9.3 Hollywood Florida

Hollywood	
Population (2021)	155,689
Median age	41.6 years
Race	68% White
Average home price	$347,000
Range of home prices	$44,000 – $7.5 million
Average elevation	3 feet

9.4 Punta Gorda

The fourth example of a community at risk is Punta Gorda, is a city of just over 21,000 people in Charlotte County (World Population Review 2021). As with many other Florida settlements, the area that became Punta Gorda was opened for development when it became a stop on the Florida Southern Railway in 1886. Soon after, phosphate was discovered on the banks of the Peace River and mining became the primary industry of the area. A trickle of residential development began when a bridge connecting Punta Gorda and the Charlotte Harbor was completed in 1921, but by 1960 its population was still only 3157. Population growth proceeded steadily since then, particularly with the opening of Punta Gorda Isles. Unlike Hollywood but more like Cape Coral, this development was designed to convert an undeveloped wetland into valuable waterfront property for single family homes. The developers Wilber Code, Alfred Johns, Sam Burchers and Robert Barbee bought 550 acres in 1957 and "set about turning their initial purchase and thousands of more acres purchased later into two prestigious waterfront communities that drew in people from all over the United States and indeed the world, forever changing the sleepy little town of about 3,000 into a magnet for retirees" (Bower 2020). They dredged 55 miles of canals 100 feet wide and 17 feet deep, using the sand to build up sites for new residences. They sold a lifestyle particularly to retirees, a "unique neighborhood laced with salt-water canals" (Williams 2011) where people could keep their boats in their backyards and walk to the golf-course.

At present more than half of its population are "seniors" and the median age is 66.5. It is more than 95% white, with an 81.5% homeownership rate. Only 13% of the Punta Gorda population was born in Florida, and 79% were born elsewhere in the United States. As of February 11, 2021, realtor.com listed 357 homes for sale in Punta Gorda. They range in price from $35,000 to $2.5 million (Table 9.4). Redfin classifies Punta Gorda as a "somewhat competitive" market in which some homes

Table 9.4 Punta Gorda

Punta Gorda	
Population (2021)	21,079
Median age	66.5 years
Race	95% White
Average home price	$278,500
Range of home prices	$35,000 – $2.5 million
Average elevation	6 feet

get multiple offers, the average home sells for 2% below list price and the average time on the market is 32 days (Redfin 2021). The median sales price is $278,500.

The average elevation is 6 feet, and virtually the entire area where those canals were created is at high risk for sea-level rise related flooding (Fig. 9.5). Punta Gorda real estate agents reported that prospective buyers are avoiding low-lying areas (average answer was "somewhat frequently"), but also reported no impact at present on house prices. One said: "*Our primary target market is an ocean-access waterfront community. While concerns [about flooding] do exist, the buyers here are willing to accept that risk.*" Another offered that the desire of buyers "*for a boat in their back yard overrides their concerns about flooding*."

9.5 Summary

Coastal residents of south Florida are vulnerable to the effects of sea-level rise in part because of the decisions of developers who platted lots for communities in order to provide waterfront access. In this chapter, we have reviewed four cases where developers created what was intended to be dry waterfront land from pre-existing wetlands by dredging the wetlands and elevating the nearby shore with fill. The result was the creation of valuable "waterfront" property, that is now vulnerable to sea-level rise as well as subsidence. In all four of these communities, regardless of the original intentions of the developers, the current residents have been left with property that almost certainly will flood with current rates of sea-level rise. And more households are settling in these communities every day: the four communities combined grew by more than 55,000 between 2010 and 2019 (U.S. Census 2021).

Real estate agents in these four communities have largely similar responses to this risk (Fig. 9.6). In this chart, the vertical axis represents the mean response on a seven-point scale for each of the sample communities. On the question of whether prospective buyers avoid low-elevation properties, real estate agents in Punta Gorda

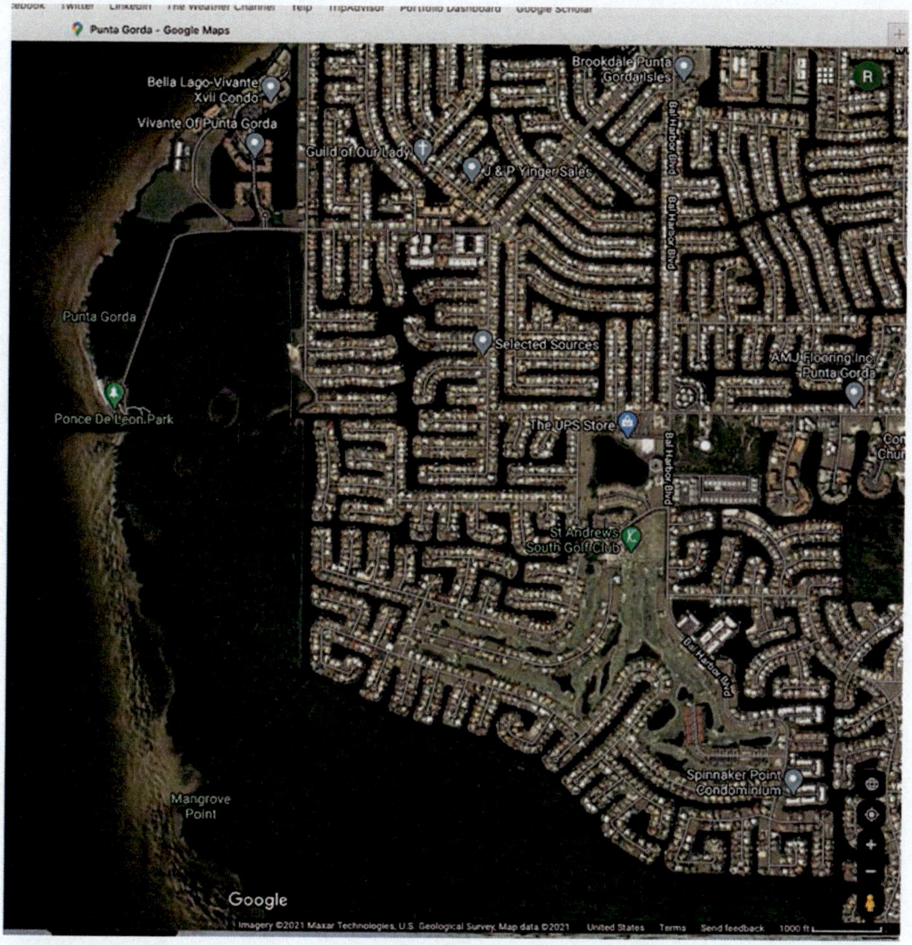

Fig. 9.5 Punta Gorda

were more likely than those in the three other areas to answer affirmatively (average score of 5.4 on a 7-point scale), but respondents in all of these communities report virtually no impact on house prices, with average scores ranging from 1.7 to 2.2. Respondents in Apollo Beach and Cape Coral were more sanguine about the property market over the next 5–10 years than those in Punta Gorda and Hollywood, but at worst expect only a "small effect" of flooding on the market. Their personal views on the reality of climate change and sea-level rise were also generally uniform, with most believing that sea-level rise and climate change is at least "probably" happening.

We have seen that prospective buyers are still very interested in purchasing waterfront property in south Florida despite the great variety and depth of information available to them about flood risk and sea-level rise. In the final chapter, we will

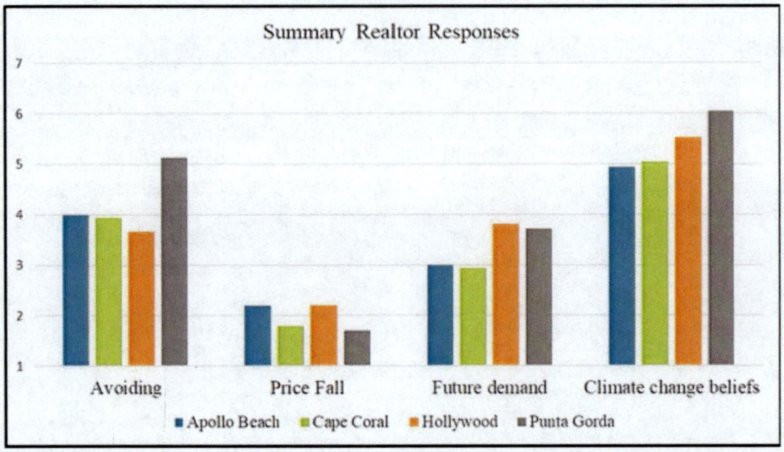

Fig. 9.6 Real estate agent responses from four vulnerable communities

discuss the implications of these empirical findings for both public policy and future research.

References

Bartling H (2008) A master-planned community as heterotopia. The villages, Florida, 165–178. In: Dehaene M, De Cauter L (eds) Heterotopia and the city. Routledge, London, pp 165–178. https://doi.org/10.4324/9780203089415

Bloetscher F, Polsky C, Bolter K, Mitsova D, Garces KP, King R, Carballo IC, Hamilton K (2016) Assessing potential impacts of sea-level rise on public health and vulnerable populations in Southeast Florida and providing a framework to improve outcomes. Sustainability 8(4):315. https://doi.org/10.3390/su8040315

Bower G (2020) Isles canal transformed the city. In: Punta Gorda Sun. Available at: https://www.yoursun.com/puntagorda/featuredsections/isles-canals-transformed-the-city/article_b8c60ea0-726f-11e9-986d-77ae97acf5ca.html

Englander J (2020) Florida is not going under water. John's Blog/Newsletter: Sea-Level Rise Now. https://johnenglander.net/florida-is-not-going-underwater-anytime-soon/

Enterprise Florida (2017) Florida: gateway to Latin America and the Caribbean. Orlando Available at: https://www.enterpriseflorida.com/wp-content/uploads/brief-florida-gateway-latin-america-caribbean.pdf

Fairfield JD (2018) The city beautiful movement, 1890–1920. Oxford Research Encyclopedia of American History. https://doi.org/10.1093/acrefore/9780199329175.013.558

Fajardo L (2016) How Miami became the capital of affluent Latin America. In: BBC News. Available at: https://www.bbc.com/news/world-us-canada-36281648

Grunwald M (2017) The boomtown that shouldn't exist. In: Politico. Available at: https://www.politico.com/magazine/story/2017/10/20/fastest-growing-city-america-florida-cape-coral-215724/

Gwyther G (2006) Paradise planned: community formation and the master planned estate. Urban Policy Res 23(1):57–72. https://doi.org/10.1080/08111140420003353304

Maschek AJ (2008) A piece of history: Apollo Beach. Hillsborough Community College Digital Library, Tampa. http://purl.flvc.org/hccfl/fd/HCC0101AB006

Mickelson J (2015) Florida's Hollywood: history and people. https://joanmickelsonphd.wordpress.com/page/2/

Mormino GR (2005) Land of sunshine, state of dreams: a social history of modern Florida. University Press of Florida, Gainesville

Nealen MJ (n.d.) Apollo Beach Florida history. https://www.maryjonealen.com/apollo-beach-florida-history

Osprey Observer (2018) History of Apollo Beach. https://www.ospreyobserver.com/local-history/history-of-apollo-beach/

Redfin (2021) Apollo Beach homes for sale. Available at: https://www.redfin.com/city/21199/FL/Apollo-Beach/housing-market

Smith C (2009) The plan of Chicago: Daniel Burnham and the remaking of the American city. University of Chicago Press, Chicago

TenEick VE (1989) History of Hollywood (1920 to 1950). Patricia M. Smith and Florida Classics Library, Hobe Sound

US Census (2019) American Community Survey, 2019

US Census Bureau (2021) City and town population totals: 2010–2019. Available at https://www.census.gov/data/tables/time-series/demo/popest/2010s-total-cities-and-towns.html

Williams L (2011) Young lions of PGI turn sand flat into home sites. http://www.lindseywilliams.org/index.htm?Articles/Punta_Gorda_Isles.htm~mainFrame

Williams CA (2017) No permit required: developing Cape Coral in 1958. In: News Press. Available at: https://www.news-press.com/story/news/local/cape-coral/2017/04/19/cape-coral-1958-development-mangroves-leonard-rosen-cynthia-williams/100584564/

World Population Review (2021) https://worldpopulationreview.com/

Chapter 10
Implications of Housing Market Response to Sea-Level Rise in South Florida

> *Living on a beach will almost feel like you are on a never-ending holiday. Beautiful sandy beaches, the clear blue sea, the gentle breeze in your hair – making that your everyday life is totally worth some spontaneous risk!*
>
> *Best Movers Florida (2021) Pros and cons of buying beachfront property. https://bestmoversinflorida.com/pros-and-cons-of-buying-beachfront-property/*

Abstract Several influential studies have suggested that house prices in coastal Florida, in areas prone to flooding associated with sea-level rise, have been stagnating or falling in comparison with price trajectories of less vulnerable locations. However, when surveyed in the late months of 2020, real estate agents reported that they have not noticed any weakening in demand for coastal properties, see no evidence of price effects from buyers, appraisers or mortgage lenders, have generally optimistic views about the future of the housing market, and expect houses even in vulnerable areas to continue selling briskly. Numerous adaptations to sea-level rise are possible, but all of them are costly and will take commitment from government entities and the general public. What remains unresolved is the issue of who should pay for these adaptations and how much will the costs of occupancy of vulnerable areas continue to be shouldered by local, state, and federal taxpayers. These are important questions not only for Florida or other coastal states, but also for the entire nation.

Keywords Florida real estate · House price trends · Adaptation

A growing body of research, as discussed in Chap. 3, has demonstrated that sales prices for housing susceptible to flooding associated with sea-level rise are beginning to stagnate or decline relative to housing in less vulnerable areas. In 2016, Zillow predicted that by 2100 one in eight homes in Florida will be underwater, wiping out over $413 billion dollars in property (Bergman 2019). These studies, understandably, have received a great deal of attention in the popular media given their massive financial implications for the local, national and even international

© The Author(s), under exclusive license to Springer Nature Switzerland AG 2022

R. Palm, T. Bolsen, *Housing Market Response to Sea-Level Rise in Florida*, Coastal Research Library 37, https://doi.org/10.1007/978-3-030-88435-2_10

market. Billions of dollars are at stake for homeowners, prospective buyers, investors and the taxpayers who subsidize the costs of living in these vulnerable locations.

The purpose of our book was to investigate another perspective on trends in the coastal Florida housing market, to assess the extent to which "professional experts" on the ground who actually sell homes in coastal South Florida have noticed such a response from buyers, or a "price discount." Its purpose was also to ascertain whether real estate agents have observed buyers paying attention to newly available information about flood-risk in the purchasing process.

To study these crucially important questions about recent dynamics within the south Florida housing market, we implemented a survey of brokers and sales associates in 13 coastal counties across six primary geographic submarkets in late 2020. As reviewed in Chap. 6, we obtained 680 high-quality responses to an IRB-approved email-based survey we administered through Qualtrics. The survey included both traditional and open-ended questions that allowed respondents to elaborate about their personal views, enabling us to provide an in-depth account of real estate agents' beliefs about the impact of new information that has become available to buyers on websites such as FloodFactor.com, and their perceptions about the behavior of other key actors in the home-purchasing process. The targeted sample of real estate agents we surveyed are in a unique position to speak about the south Florida housing market given their expertise, and the large number of homes they sell. The sample we obtained was large and diverse across the geographic submarkets selected for the study. Notwithstanding a predictably low response rate that is common for email-based sampling approaches, we effectively recruited a sample of experts in the real estate industry in south Florida whose experience with the market is worth noting.

10.1 Key Findings

Conventional wisdom suggests that concern with flood risk associated with sea-level rise has had an impact on house prices in Florida. Our findings do not support this assertion. Instead, we found that almost no one selling homes in south Florida believes that there has been an impact of sea-level rise or flood-risk concern on coastal-home sales prices. As we reported in Chap. 8, only 11 out of 680 of our respondents perceived this as happening "very frequently," while the vast majority of professional experts simply do not see any "price discounting effect." They also reported that *mortgage lenders* are rarely, if ever, declining or charging more points for loan applications in low-lying coastal areas that are flood prone. These findings were consistent across the six housing submarkets we studied. The brokers and sales associates also reported that *home appraisers* generally do not consider flood-risk due to sea-level rise as a key factor in determining the assessed value of coastal homes in South Florida, although the cost of flood insurance is a factor that buyers often consider when making a purchase in low-lying elevation locations (see Chap. 7).

We also asked real estate agents for their views about the likely trajectory of the residential property market in low-lying coastal areas over the next 5–10 years, and if they think it is likely there will be lower demand or lower home sale prices over this period due to flooding concerns. Although there was some variance based on geographic submarket, with higher levels of concern in Miami-Dade County compared to other areas, most respondents selling homes across south Florida believe that house prices in coastal flood-prone areas will not be affected by the prospect of sea-level rise in the near future. They note that insurance premiums may increase in vulnerable areas but predict that homebuyers will simply absorb these costs. If these perceptions are accurate and reflective of what is actually happening "on the ground" in south Florida, they raise questions about the evidence showing that concern about sea-level rise and coastal flooding are already playing a powerful role in shaping the behavior of prospective coastal homebuyers.

Although the real estate agents we surveyed largely do not perceive any current or near-term future "price discount" for flood-risk coastal properties, as detailed in Chap. 8, the majority do support providing prospective buyers with the most up-to-date and accurate information about current and future flood risks associated with properties they sell. For example, as reported in Chap. 7, a majority of real estate agents expressed support for the NAR's decision to add "Flood Factor" information to property listings on its website, while only a small minority expressed reservations and/or said that they would be unlikely to recommend the information to a potential buyer. However, they also reported that prospective sellers and buyers are *not paying attention* to this new information about flood-risk associated with future sea-level rise in the sales of coastal properties in South Florida. Other factors in the home-purchasing decision appear to outweigh any considerations about long-term risks stemming from flooding resulting from sea-level rise, and this may be exacerbated by the fact that much of the risk is "externalized" through artificially low insurance premiums and the eagerness of mortgage lenders to make "conforming" loans.

In Chap. 4, we reviewed how individuals' beliefs about climate change and sea-level rise have been powerfully shaped by ideological polarization in the U.-S. Liberal Democrats overwhelmingly accept the scientific consensus and see climate change and sea-level rise as major threats, whereas conservative Republicans tend to be more skeptical about the science and less concerned. These deep partisan divides have been documented among homeowners and residents in coastal communities of South Florida who live in areas at substantial risk of flood damage (Palm and Bolsen 2020). In addition, variation in the public's beliefs about climate change and sea-level rise, often measured at the county-level, has been linked in other research with the existence of a "price discounting" effect in several large studies of buyer behavior in South Florida, as reviewed in Chap. 3. This research also suggests that price discounting in low-lying coastal areas is more likely where concern about climate change and sea-level rise for the average resident is relatively higher. Given the theorized importance of these *individual-level beliefs* in shaping home buyer behavior, we included questions at the end of our survey asking real estate agents about their own views about climate change and sea-level rise, both to provide a snapshot of this group's views on these matters and to evaluate the degree to which variance in their beliefs on these issues was predictive of their attitudes about home buyer behavior and the existence of a "price discount" in areas where they sell properties.

Most real estate agents we surveyed believe that climate change and sea-level rise are happening and interrelated; however, differences in the strength of these views were often correlated with variance in our key outcome measures related to home buyer behavior and price discounting effects. As reported in Chaps. 7 and 8, **stronger belief in climate change and sea-level rise was associated with** (1) an increased likelihood of responding that buyers are avoiding low-elevation coastal properties, (2) higher levels of support for the NAR decision to link to FloodFactor.com on its website, (3) an increased willingness to recommend FloodFactor.com to potential buyers, (4) an increased perception that lenders and appraisers are considering elevation, and (5) an increased likelihood of stating that future demand in low-lying coastal areas will suffer as a result of sea-level rise. In sum, differences in real estate agents' views about climate change and sea-level rise are associated with differences in their views about aspects of the current and future housing market in South Florida. Further, as reported in Appendix 3, party identification is the strongest predictor of variance in real estate agents' beliefs about climate change and sea-level rise.

10.2 Looking to the Future

Scientists project that sea-levels will continue to rise and that this pace will occur at a much faster rate compared to historical trends (Fig. 10.1). As this occurs, extreme weather events, such as intense hurricanes and flood damage due to storm surge, will occur with greater frequency. Risk associated with flood damage currently presents a substantial threat to an estimated 906,465 residences in the state of Florida, and related damages are expected generate a collective financial loss of almost 8 billion dollars in 2021 (First Street Foundation 2021). Those asset losses are projected to increase by 77% over the next 30 years in Florida and threaten an additional 67,069 homes currently not projected to be at substantial risk of flood damage.

By 2060, some climate models predict two feet or more of sea-level rise in counties in South Florida such as Miami-Dade. In theory, Floridians can choose among four strategies: "protect, accommodate, retreat and avoid" (Doberstein et al.

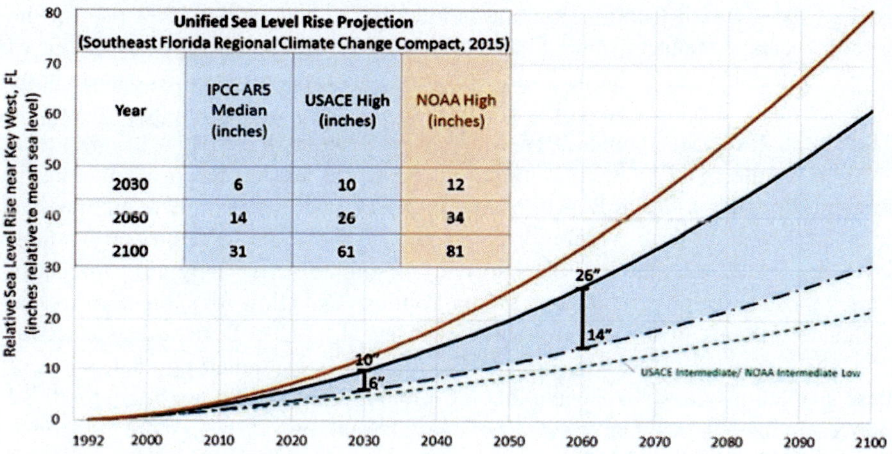

Fig. 10.1 Projections of rising sea-levels compared to historical trends. (Source: Southeast Florida Climate Compact)

2019). Protection involves engineering solutions such as the construction of sea walls, levees or surge barriers, the building up of sand dunes or beach renourishment, raising roads, building stormwater pumps, or slowing land subsidence through the injection of purified wastewater into the ground (Sealevelrise.org n.d.). Accommodating to sea-level rise could include elevating housing and using flood resistant construction materials (Mclean and Watson 2009). Retreat from the flood-prone area could be initiated by government acquisition of vulnerable property and even the construction of floating neighborhoods or cities to accommodate the displaced population (Al 2018). Avoidance includes zoning restrictions to prevent further development in vulnerable areas. Local and state governments in Florida have adopted a smattering of these strategies (Florida Department of Economic Opportunity n.d.).

There are several examples of local counties and municipalities adopting protection or engineering solutions such as elevating homes, raising roads, building storm walls, and improving drainage systems (Scott et al. 2020). Of Palm Beach County's 47 beaches, 33 have been classified as "critically eroded", and its neighboring communities face similar problems with their own beaches, which led the State Legislature to provide $50 million dollars annually to replenish sand on disappearing beaches (Carroll 2021). Officials in Miami Beach recently released a $206 million dollar plan to overhaul an "antiquated drainage system" that seeks to address rising sea-levels with "more pumps, wells to store storm runoff, higher sea walls and 'back flow' preventers for drainpipes flowing into Biscayne Bay" (Morgan 2015). This plan was designed, in part, to ensure developers and buyers are reassured about waterfront real estate values (Flavelle and Mazzei 2021). The plan must be approved by the Miami Beach Commission and is intended to address sea-level rise for the next 20 years. "Managed retreat," or "the purposeful, coordinated movement of people and assets out of harm's way" (Siders 2019) has received recent attention from scientists and local governments, but has not been widely adopted in Florida (Hino et al. 2017; Mach et al. 2019; Mast 2019; Scott et al. 2020).

Some of the most expensive houses in the world are located in south Florida communities such as Palm Beach (Marcoux 2021), and therefore there are massive financial incentives to find "solutions" to sea-level rise to preserve property values. Some argue that this has led many to deny or minimize the long-term threats of sea-level rise and climate change in favor of options that allow for short-term coastal resilience instead of coastal retreat (Scott et al. 2020; Defries et al. 2019; Oppenheimer et al. 2019; Ruppert and Grimm 2013). Although popular, some of these adaptations are so expensive that most communities could not begin to afford to invest in them, leaving large swaths of coastal South Florida even more vulnerable (Carroll 2021; Bergman 2019).

In short, scientific projections of sea-level rise and recent experience with flood damage have prompted sporadic policy responses from local communities across South Florida. Nonetheless, households continue to seek housing in areas that are likely to be underwater in the future. The short-term benefits of living on the coast and the amenities that accompany a beach lifestyle appear to factor more than the long-term prospect of sea-level rise in buyers' decisions about where to live and what to pay for a home.

Researchers studying housing market response in flood-prone areas like south Florida should consider the perspective of real estate professionals. For example, it will be important to account for the differences in estimates of house price trends and consumer demand reported by those selling homes in south Florida as opposed to calculations based on large-scale quantitative analyses. It is of course possible that real estate agents' views are "biased" by their own self-interest, and they are "blind" to the more subtle "macro-level" shifts in buyer behavior in the areas where they sell properties. On the other hand, it is also possible that the existing studies that demonstrate an "elevation-related" price discount in areas along the coast of south Florida fail to account for variables that are affecting buyer behavior, leading to unwarranted conclusions that fit a popular narrative about concerns over climate-induced sea-level rise playing a significant role in the sale of coastal homes.

Looking to the future, the observations of south Florida real estate agents suggest a continued strong demand for vulnerable coastal homes with little or no impact, at least to date, on house price. A rise in costs for mandatory flood insurance has no impact on property owners without a mortgage loan. And for the very wealthy, the incursion of the ocean can be fended off for a few decades with beach renourishment or the construction of sea walls by the city, county, or state. The majority of homeowners, however, have mortgage financing from federally insured lenders and will have to bear the increased costs if future revisions to flood insurance policy put more financial burden on those at greatest risk.

The rising costs of protecting and insuring low-elevation, flood-prone coastal homes raise numerous thorny policy questions. It is estimated that 4.3 million homes across the U.S. are currently at substantial risk of flooding, and the amount of estimated damage that will occur to these properties is expected to increase substantially as a result of the changing climate (First Street Foundation 2021). Who should pay for the minimization of risk and the true costs of living on the coast? How much risk should homeowners versus insurers share in terms of the burden of rising costs for insuring coastal properties? How will insurance costs and coastal adaptation policies affect homeowners in different income brackets who live in these places? How much of the cost burden for coastal adaptation should be shouldered by local, state, and federal taxpayers? As a nation, how do we choose between the painful options of defending these coasts with increasingly costly engineering works or watching as more property is lost to the sea. Or, as a nation should we plan for managed retreat? These are looming questions that must be addressed not only in South Florida but across coastal communities in the United States.

Rising sea-levels resulting from of climate change are now inevitable. Many of the houses currently being sold in south Florida will not out-last their 30-year mortgages, and the re-sale of houses vulnerable to sea-level rise is very likely to become increasingly difficult. Florida policymakers to date have either ignored the risk or have taken limited measures to patch particular weaknesses, sometimes increasing the risk to other parts of the state. And warnings from organizations outlining the likely impacts of rising sea-levels on beachfront property seem not to have been heeded. One would like to believe that "the market" will take care of this issue in that homebuyers, recognizing the looming risks, will discount prices on vulnerable properties, eventually reducing their attractiveness and value: that with more detailed information about the risk of flooding, the rational home buyer would respond by avoiding risky areas or, at the very least, insisting on a price discount. But the testimony of Florida real estate agents we surveyed casts doubt on the assumption that the "market" has currently integrated this risk.

What is needed in the near future is a national assessment of the impacts of climate change that includes realistic portrayals of the way in which buyers are making decisions about houses to purchase, the way that mortgage lenders are making decisions about the value and terms of non-conforming mortgage loans and off-load the risks to conforming loans, and the way that property appraisers are assessing the current and future value of individual properties. The goal should be the identification of systematic and concrete policies that could be implemented by both the public and private sectors to improve the safety and security of all Americans. Such policies could include systematic ways to manage retreat from vulnerable environments, and their conversion to other uses than as sites for private residences. It is not too much to expect that we as a nation develop a strategy not only to curb climate change itself, but also to keep Americans safe now and for the next century as climate change continues to play out. To do so, we will need realistic assessments of how the market is actually working so that we can make recommendations on policy measures that will be effective for the long-term.

References

Al S (2018) Adapting cities to sea level rise: green and gray strategies. Island Press, Washington, DC. https://doi.org/10.5822/978-1-61091-908-1

Bergman MM (2019) Florida is drowning. Condos are still being built. Can't humans see the writing on the wall? In: The Guardian. Available at: https://www.theguardian.com/environment/2019/feb/15/florida-climate-change-coastal-real estate-rising-seas

Best Movers Florida (2021) Pros and cons of buying beachfront property. Available at: https://bestmoversinflorida.com/pros-and-cons-of-buying-beachfront-property/

Carroll J (2021) The damage in Florida from rising sea levels already is here | Commentary. In: Orlando Sentinel 8 Feb 2021. Available at: https://www.orlandosentinel.com/opinion/guest-commentary/os-op-florida-sea-level-rise-damage-here-20210208-3mibpyp6ivhclacigj5pnycqgq-story.html. Accessed 10 Feb 2021

DeFries RS., Edenhofer O, Halliday AN, Heal G M, Lenton T, Puma M, Rising J, Rockström J, Ruane AC, Schellnhuber JJ, Stainforth D, Stern N, Tedesco M, Ward B (2019) The missing economic risks in assessments of climate change impacts. Grantham Research Institute on Climate Change and the Environment. Available at: https://www.lse.ac.uk/granthaminstitute/publication/the-missing-economic-risks-in-assessments-of-climate-change-impacts/

Doberstein B, Fitzgibbons J, Mitchell C (2019) Protect, accommodate, retreat or avoid (PARA): Canadian community options for flood disaster risk reduction and flood resilience. Nat Hazards 98(1):31–50. https://doi.org/10.1007/s11069-018-3529-z

First Street Foundation (2021) The cost of climate. America's growing flood risk. Available at: https://assets.firststreet.org/uploads/2021/02/The_Cost_of_Climate_FSF20210219-1.pdf

Flavelle C, Mazzei P (2021) Miami says it can adapt to rising seas. Not everyone is convinced. 2 Mar 2021. In: New York Times. Available at: https://www.nytimes.com/2021/03/02/climate/miami-sea-level-rise.html

Florida Department of Economic Opportunity (n.d.) How countries, states, and Florida address sea level rise: a compendium of climate adaptation research. Available at: https://content.sierraclub.org/grassrootsnetwork/sites/content.sierraclub.org.activistnetwork/files/teams/documents/CompendiumNationalStateLocalAdaptationProjects.pdf

Hino M, Field CB, Mach KJ (2017) Managed retreat as a response to natural hazard risk. Nat Clim Change 7:364–370. https://doi.org/10.1038/nclimate3252

Mach KJ, Kraan CM, Hina M, Siders AR, Johnston EM, Field CB (2019) Managed retreat through voluntary buyouts of flood-prone properties. Sci Adv 5(10):eaax 8995. https://doi.org/10.1126/sciadv.aax8995

Marcoux S (2021) Take a peek at 25 of the most expensive houses in the world. In: Veranda. Available at: https://www.veranda.com/decorating-ideas/g27550202/most-expensive-house-in-the-world/

Mast T (2019) Bonita springs might buy homes in neighborhoods flooded by Hurricane Irma. In: Naples Daily News. Available at: https://www.naplesnews.com/story/news/local/communities/the-banner/2019/07/24/bonita-springs-might-buy-imperial-river-properties-flooded-during-hurricane-irma/1816564001/

Mclean S, Watson P (2009) A practical approach to development of housing on floodplain land in the UK. J Build Apprais 4(4):311–320. https://doi.org/10.1057/jba.2009.10

Morgan C (2015) Rising sea comes at a cost for South Florida cities. Coastal Systems International. https://coastalsystemsint.com/media/Rising_sea_comes_at_a_cost_for_South_Florida_cities.html

Oppenheimer M, Oreskes N, Jamieson D, Brysse K, O'Reilly J, Shindell M, Wazeck M (2019) Discerning experts: the practices of scientific assessment for environmental policy. University of Chicago Press, Chicago. https://doi.org/10.7208/chicago/9780226602158.001.0001

Palm R, Bolsen T (2020) Results from south Florida experiment. In Climate change and sea level rise in South Florida, p81-92. Springer Nature Switzerland AG, Cham, doi:https://doi.org/10.1007/978-3-030-32602-9_6

Ruppert T, Grimm C (2013) Drowning in place: Local government costs and liabilities for flooding due to sea-level rise. Fla Bar J 87(9):29–. https://www.floridabar.org/the-florida-bar-journal/drowning-in-place-local-government-costs-and-liabilities-for-flooding-due-to-sea-level-rise/

Scott M, Lennon M, Tubridy D, Marchman P, Siders AR, Main KL, Hermann, V Mark, Mick Lennon, Daniel Tubridy, Patrick Marchman, A. R. Siders, Kelly Leilani Main, Victoria Herrmann, Butler D, Frank K, Bosomworth K, Blanchi R, Johnson C (2020) Climate disruption and planning: resistance or retreat? Plan Theory Pract 21(1):125-154. doi:https://doi.org/10.1080/14649357.2020.1704130

Sealevelrise.org (n.d.) Solutions can protect coastal communities. https://sealevelrise.org/solutions/

Siders AR (2019) Managed retreat in the United States. One Earth 1(2):216–225. https://doi.org/10.1016/j.oneear.2019.09.008

Appendices

Appendix 1: Questionnaire

How long have you been a real estate professional?

1. Less than a year
2. 1–2 years
3. 3–5 years
4. 6–10 years
5. 11–20 years
6. More than 20 years

Are you a broker or a sales associate?

1. Broker
2. Sales associate

Approximately how many houses do you sell in an average year?

1. 0
2. 1–2
3. 3–5
4. 6 or more

How long have you lived in Florida?

1. Less than a year
2. 1–2 years
3. 3–5 years
4. 6–10 years
5. 11–20 years
6. More than 20 years

© The Author(s), under exclusive license to Springer Nature Switzerland AG 2022
R. Palm, T. Bolsen, *Housing Market Response to Sea-Level Rise in Florida*, Coastal Research Library 37, https://doi.org/10.1007/978-3-030-88435-2

How old are you?

1. Under 18
2. 18–24
3. 25–34
4. 35–44
5. 45–54
6. 55–64
7. 65–74
8. 75–84
9. 85 or older
10. Prefer not to answer

What is your sex?

1. Male
2. Female
3. Prefer not to answer

Which of the following do you consider to be your primary racial or ethnic group?

1. White
2. African American
3. Asian American
4. Hispanic
5. Other
6. Prefer not to answer

What is the highest level of education you have completed?

1. Less than high school
2. High school graduate
3. Some college
4. 2-year degree
5. 4-year degree
6. Professional degree
7. Doctorate

Generally speaking, which of the options on the scale below best describes your party identification?

1. Strong Republican
2. Weak Republican
3. Lean Republican
4. Independent
5. Lean Democrat
6. Weak Democrat
7. Strong Democrat

In your experience, how important are the following factors for buyers when deciding on a neighborhood in which to purchase a house?

	Not at all important (1)	A little important (2)	Somewhat important (3)	Moderately important (4)	Very important (5)	Extremely important (6)	Most important (7)
Age of the neighborhood							
Noise levels in the neighborhood							
Parking/traffic in the neighborhood							
Presence of children in the neighborhood							
Proximity to a beach, park or playground							
Proximity to work/commute time							
Quality of the local public schools							
Safety from flooding in the neighborhood							
Proximity to restaurants, shops, and entertainment							

Please list any additional factors.

To what extent do you believe that buyers are avoiding property in low-elevation coastal areas because of possible flooding?

1. Not at all
2. Rarely
3. A little
4. Somewhat
5. Somewhat frequently
6. Very frequently
7. All the time

Do you think buyers are avoiding properties at low elevations? Why or Why not?

If some buyers are avoiding low-elevation coastal areas, this would reduce overall demand and therefore reduce prices. To what extent have you observed that house prices are either falling or not rising as rapidly for properties at risk of flooding?

1. Not at all
2. Rarely
3. A little
4. Somewhat
5. Somewhat frequently
6. Very frequently
7. Always

We might expect that buyers who are planning to live in the property year-round themselves might respond differently to possible flooding than those investing in property or buying it as a vacation home. In your opinion, do different types of buyers avoid properties in low-lying coastal areas? In your opinion, do different types of buyers avoid properties in low-lying coastal areas?

	Not at all (1)	Rarely (2)	A little (3)	Somewhat (4)	Somewhat frequently (5)	Very frequently (6)	Always (7)
Owner-occupiers							
Investors							
Second homebuyers							

Have you observed mortgage lenders declining loan applications or increasing charges for loans in low-lying areas that are flood prone, in the form of points or mortgage insurance, for example?

1. Not at all
2. Rarely
3. A little
4. Somewhat
5. Somewhat frequently
6. Very frequently
7. All the time

Appraisers might also consider elevation or flood likelihood in their valuation of property. To what extent do you believe that property appraisers consider elevation or the likelihood of flooding in their assessment of home value?

1. Not at all
2. Rarely
3. A little
4. Somewhat
5. Somewhat frequently

6. Very frequently
7. All the time

Does elevation matter more or less for homes that are at the high end of the market as opposed to those selling for lower prices?

1. Elevation matters much more at the high end of the market
2. Elevation matters somewhat more at the high end of the market
3. There are no differences between the high end and the low end with respect to elevation
4. Elevation matters somewhat more at the low end of the market
5. Elevation matters much more at the low end of the market.

In August of this year, the website Realtor.com began to put a new category of information called "Flood Factor" as part of the details of property listings. This new addition includes detailed information on the projected risks of future flooding that may threaten a property.

Have you seen any response to this new information from buyers or sellers?

If yes, has the response from buyers overall been positive or negative?

1. Extremely negative
2. Moderately negative
3. Slightly negative
4. Neither positive nor negative
5. Slightly positive
6. Moderately positive
7. Extremely positive

If you have seen any response from sellers, has it been positive or negative?

1. Extremely negative
2. Moderately negative
3. Slightly negative
4. Neither positive nor negative
5. Slightly positive
6. Moderately positive
7. Extremely positive

What is your view of the decision of the National Association of Realtors® to add this "Flood Factor" information to property listings?

1. Strongly oppose
2. Oppose
3. Somewhat oppose
4. Neither support nor oppose
5. Somewhat support
6. Support
7. Strongly support

In a few sentences, please give us your opinion about the addition to the listings of these maps and the information that accompanies them.

How likely are you to recommend that potential buyers view this new "Flood Factor" information included on property listings?

1. Extremely unlikely
2. Moderately unlikely
3. Slightly unlikely
4. Neither likely nor unlikely
5. Slightly likely
6. Moderately likely
7. Extremely likely

How do you see the near-term future for residential property in low elevation coastal areas? Will the residential property market see less demand and lower prices in areas prone to flooding in the next 5–10 years?

1. No effect of flooding
2. Almost no effect of flooding
3. Very small effect of flooding
4. Small effect of flooding
5. Moderate effect of flooding
6. Large effect of flooding
7. Very large effect flooding

To what extent do you believe that the sea-level is rising?

1. Definitely is NOT rising
2. Very likely is NOT rising
3. Probably is NOT rising
4. Not sure
5. Probably IS rising
6. Very likely IS rising
7. Definitely IS rising

Do you think that climate change is happening?

1. Definitely is NOT happening
2. Very likely is NOT happening
3. Probably is NOT happening
4. Not sure
5. Probably is happening
6. Very likely is happening
7. Definitely is happening

To what extent do you think sea-level rise is related to climate change?

1. Definitely is NOT related to climate change
2. Very likely is NOT related to climate change

3. Probably is NOT related to climate change
4. Not sure
5. Probably IS related to climate change
6. Very likely IS related to climate change
7. Definitely IS related to climate change

Please describe your view of the future (over the next 5–10 years) for residential real estate in Florida.

Specifically, do you think that sea-level rise will weaken demand for coastal property in your market area? Why?

And do you think that sea-level rise will result in lower prices for coastal property in your market area? Why?

Appendix 2: Survey Experiment

We embedded a randomized experiment using the information available for all listings now on the realtor.com website. Although all the real estate agents have access to the full set of information including the text and maps, we didn't know how familiar our respondents would have been with this addition to the property listing that went into effect in the summer of 2020. We therefore wanted to know whether placing the full information vs. just the text about the listing in our questionnaire would elicit a different response from real estate agents.

All respondents were presented with the following text: "In August of this year, the website Realtor.com began to put a new category of information called 'Flood Factor' as part of the details of property listings. This new addition includes the estimated FEMA zone and an estimate of the chance of flooding at least once over the next 30 years."

We randomly assigned half of the respondents to a "treatment" that included a visual depiction of the property listing as it appears in the realtor.com website. This depiction includes a photo and some general information, as well as indicating its FEMA zone and its Flood Factor "score" (i.e., "7" on the 1–10 scale). We also included the map showing projected flooding and depth of flooding that could be associated with the property listing. This additional language was presented to respondents in the treatment condition: "Here is an example of a real estate listing with the new "Flood Factor" details provided. This 'Flood Factor' information includes this detailed map showing the projected future flooding for this property portraying the likelihood of flooding of a given depth reaching the largest building on the property."

We asked the two groups whether they had noticed any response to the website information from buyers or sellers, if they supported the National Association of Realtors® (NAR) decision to integrate this information into its property listings, and the likelihood of recommend it to future homebuyers. We compared the responses of those exposed to either the "text-only" or text plus "visual treatment" and found no statistically significant differences between these groups (Fig. A.1). Most respondents in each group reported that they had not seen any response from either buyers or sellers.

The treatment also had no influence on real estate agents' opinion about the decision of the NAR to add the "Flood Factor" information to property listings. There was also no significant difference between the groups on willingness to recommend that future buyers view the property listing and information (4.77 versus 4.73 mean scores for the control and treatment group respondents on this item) (Table A.1).

Several factors may account for the absence of any significant differences in response to the two versions of the questionnaire. First, as noted, most respondents

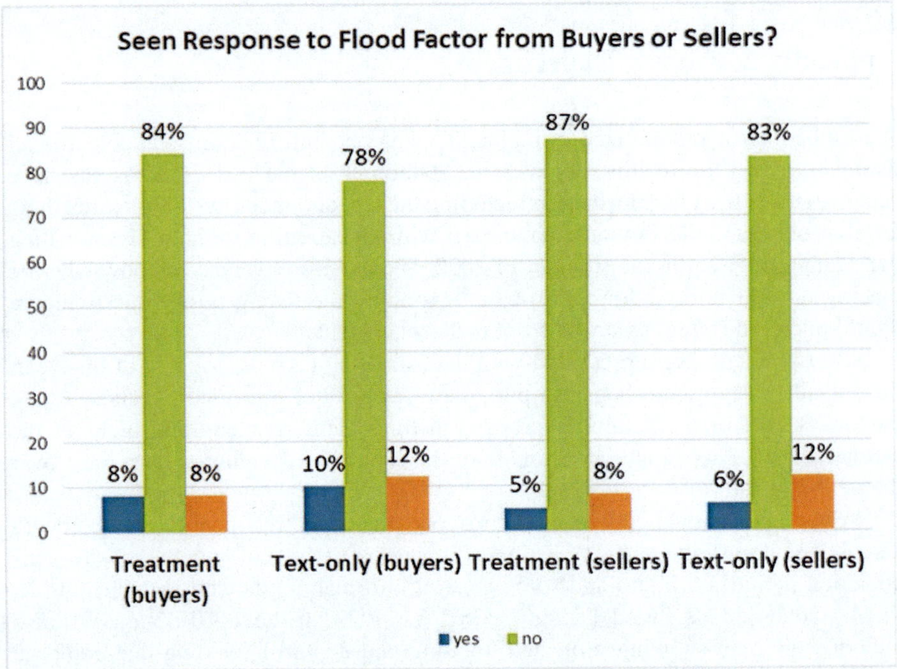

Fig. A.1 Response to flood factor by treatment

Table A.1 Difference in support for NAS decision and likelihood of recommending between control vs. treatment

Variable	Control	Treatment	Difference	p-value
Decision	4.78	4.91	−0.13	0.277
Recommend Info	4.77	4.73	0.04	0.774

Note: Two-tailed t-tests and corresponding p-values
* p<0.10, ** p<0.05, *** p<0.01

reported no response to the new information that was released, resulting in a highly skewed distribution on the outcome measure across conditions. Second, although there was greater variance in respondents' views about the decision of the NAR and the extent to different respondents were willing to recommend this information to buyers, the "treatment" in the experiment was relatively subtle and only included a single "example listing" from a property likely unfamiliar to most respondents. Third, most of the respondents in our sample were likely to have already been aware that this information was available to their buyers, and therefore had already formed an opinion about the usefulness of this information and the NAR's decision. It thus may not be surprising that in the context of this survey a subtle difference in how this information was presented to respondents had no impact on their view.

Appendix 3: Predictors of Views About SLR and Climate Change

As reviewed in Chap. 5, we asked three questions to elicit the opinions of the real estate agents on climate change and sea-level rise. The first was "to what extent do you believe that the sea-level is rising" with possible responses ranging from (1) "definitely is not rising" to (7) "definitely is rising", with (4) being "not sure". The second question was "do you think that climate change is happening?" and the third was "to what extent do you think sea-level rise is related to climate change?". Responses to these three questions were highly inter-correlated (Cronbach's alpha = 0.94), and therefore we created a single scaled variable to evaluate the views of real estate agents about climate change induced sea-level rise (Sea-level Scale). Table A.2 reports the results from an OLS regression of this scale on several demographic and political characteristics. Not surprisingly, we find that the party identification of individual real estate agents plays a dominant role in explaining variance in their views about sea-level rise and climate change. Age and education are the only other individual-level characteristics that are significantly associated with variance in respondents' views on these topics.

Table A.2 Predictors of response to "sea-level scale"

	Coefficient	p-value
Age	−0.10*	0.022
	(0.04)	
Education	0.09*	0.030
	(0.04)	
Female	0.09	0.381
	(0.10)	
Minority	−0.09	0.504
	(0.13)	
Republican	−0.93***	0.000
	(0.12)	
Democrat	0.93***	0.000
	(0.13)	
Broker	0.05	0.655
	(0.10)	
Years lived in Florida		
−0.10	0.080	
	(0.06)	
How many houses sold		
0.00	0.965	
	(0.11)	
Atlantic coast	0.01	0.888
	(0.11)	
Constant	6.19***	0.000
	(0.50)	
N	613	
Adjusted R2	0.283	
AIC	1988.2	
BIC	2036.8	

Note: Cell entries are OLS coefficients with standard errors in parentheses below
* $p < 0.05$, ** $p < 0.01$, *** $p < 0.001$